CHEVROLET'S
HOT ONES
1955·1956·1957

Text by Anthony Young
Photography by Mike Mueller

Motorbooks International
Publishers & Wholesalers

First published in 1995 by Motorbooks International Publishers & Wholesalers, PO Box 2, 729 Prospect Avenue, Osceola, WI 54020 USA

Motorbooks International books are also available at discounts in bulk quantity for industrial or sales-promotional use. For details write to Special Sales Manager at the Publisher's address

Library of Congress Cataloging-in-Publication Data
Young, Anthony.
   Chevrolet's hot ones: 1955-1956-1957/Anthony Young, Mike Mueller.
      p. cm.
   Includes index..
   ISBN 0-87938-994-X (hard cover)
   1. Chevrolet automobile--History. I. Mueller, Mike. II. Title.
   TL215.C48Y682   1995
   629.222--dc20   95-5975

On the front cover: Chevy's triumvirate: 1955, 1956, and 1957 Bel Airs. The '55 and '57 are owned by Bill and Barbara Jacobsen of Odessa, Florida, and the '56 is owned by Dennis and Mary Schrader of Tarpon Springs, Florida.

On the title page: A first-year Bel Air with all the trimmings. This droptop beauty belongs to Larry Young of Port Charlotte, Florida.

On the back cover: Main image: Chevrolet's Nomads beautifully combined styling and function. The 1955 in the background and the '57 in the foreground both belong to Erol and Susan Tuzcu of Del Ray Beach, Florida. Second image: Chevrolet ad for 1956.

Printed in China

Contents

Curves don't come too sharp or hills too steep for this nimble new Chevy. It's sweet, smooth and sassy with new velvety V8 power, new roadability, a new ride and everything it takes to make you the relaxed master of any road you travel.

Bring on the mountains! This new Chevy takes steep grades with such an easy-going stride you hardly even give them a thought. There's horsepower aplenty tucked away under that hood—the Chevy kind, just rarin' to handle any hill you aim it at.

And no matter how curvy the road may be, a light touch keeps Chevrolet right on course. You'll like the solid way it stays put on sharp turns.

A car has to have a special kind of build to handle and ride and run like a Chevy. It has to have Chevrolet's low, wide stance, its outrigger type rear springs and well-balanced weight distribution, with the pounds in the right places. Try one soon at your Chevrolet dealer's. . . . Chevrolet Division of General Motors, Detroit 2, Michigan.

The road isn't built that can make it breathe hard!

More beautifully built and shows it—the new Chevrolet with Body by Fisher.

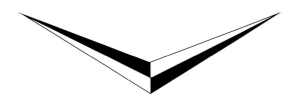

Acknowledgments

On April 15, 1955, my parents, Helen and Randy Young, were married and then left on their honeymoon in a 1955 Two-Ten. Little did I know that one day I would write about their first Chevrolet—the first of many in the years to come. This book is dedicated to them, my wife Annie, and my daughters, Erin and Katie.

I want to express my appreciation to the owners of the beautiful cars Mike Mueller has photographed for this book. They are Bill and Barbara Jacobsen of Odessa, Florida; Dennis and Mary Schrader of Tarpon Springs, Florida; Erol and Susan Tuzcu of Del Ray Beach, Florida; Walter Cutlip of Longwood, Florida; Jerome Cain, Fred Gaugh, and Terry Sheafer of Lakeland, Florida; Kevin Mueller of Rockton, Illinois; Steve and Chris Mech-

ling of Champaign, Illinois; Lloyd Brekke of Lakeland, Florida; Floyd Garrett of Fernandina Beach, Florida; and Jim Kelsey of the Klassic Auto Museum in Daytona Beach, Florida.

I am grateful to Bob Ashton of Auto Know in Utica, Michigan, for loaning me the color ads to use in this book. Jack Underwood of GM Product Information has always supported my automotive writing projects and did so with this book by supplying me with copies of the Chevrolet Features books, black and white transparencies of Zora Arkus-Duntov's assault on Pikes Peak, and copies of Chevrolet brochures. Finally, I must thank Ruth Bouldes of Chevrolet Public Relations for permission to review the priceless photo books so I could select the black and white photos needed.

Using the effortless hill-climbing theme, this ad was meant to show the power of the V-8s available in the 1957 Chevrolet. "Curves don't come too sharp or hills too steep for this nimble new Chevy," the ad reads. Some poor hapless hot rod owner (a Ford perhaps?) has to wait while his car cools off after the strain of those hills in the background. This ad appeared in *National Geographic*.

The Classic Tri-Chevys:
1955, 1956, 1957

The classic Tri-Chevys—the 1955, 1956, and 1957 models—hold an unmatched fascination for all postwar Chevrolet enthusiasts. Why these three particular years? What is it about the '55, '56, and '57 Chevys that endear them to countless bowtie enthusiasts?

There are two primary reasons: the 1954 Chevrolet and the 1958 Chevrolet. The 1954 Chevrolet still carried with it stylistic and engineering features left over from the resumption of passenger car production after World War II. The car was still powered by a straight-six-cylinder engine, while other car manufacturers already had V-8s available.

For 1955, Chevrolet rolled out totally new styling and engineering that made the car truly exciting to own and drive. And then there was that new small-block V-8. The 1956 model boasted distinctive new styling in response to the annual model change dictum. The fin era was ushered in with the 1957 model year, perhaps the favorite among the three years for collectors and hot rod enthusiasts. All three model year cars were built upon the newly engineered chassis that kept the car reasonably light and agile.

With the 1958 model year, the Chevrolet underwent a complete change, every bit as radical as the jump from 1954 to 1955. The trim lines and engineering architecture of the previous three years was abandoned in favor of longer, lower, wider, and heavier. This, in turn, was followed by the

1959 Chevrolet, the year of Fin Excess.

The years bracketing the Classic Chevys, or Tri-Chevys as they are called, have proved to be forgettable cars, while the 1955, 1956, and 1957 cars have endured and prospered. When was the last time you saw a '54 Bel Air? Or a '58 Nomad? Yes, there was a 1958 Nomad, but it was a Nomad in name only.

Central to the Tri-Chevys' popularity is their look. Each remains handsome to this day. With the 1955 model year, a new enthusiasm and optimism pervaded not just the advertising but the car itself. With the advent of the small-block V-8, the car became an object to be raced as well, and not just admired. The racing and the victories added immeasurably to the car's mass appeal, fueled by its affordability. After all, Chevrolet billed itself as "Stealing the thunder from the high-priced cars!"

GM's Motorama offered the first veiled glimpse of Chevy's "hot one." Motorama began in 1953 as an outgrowth of the annual industrialist luncheons hosted by GM Chairman Alfred P. Sloan in New York. These luncheons expanded to become the "Transportation Unlimited" shows of 1949 and 1950. The Korean War put a stop to these shows, but they reappeared in much more lavish form in 1953 as the General Motors Motorama.

Motorama became the focal point of automotive journalists eager to see current and forthcoming GM cars. Of particular interest were the dazzling concept show cars courtesy of Harley J. Earl's Advanced Styling Section within the GM Styling Section.

With knowledge of the upcoming 1953 Motorama, Earl used all the resources at his disposal to "wow" the show goers when the show opened. Present were the Cadillac Orleans and LeMans, Buick's Wildcat, Pontiac's Landau, and Oldsmobile's Starfire. No less a crowd pleaser was Chevrolet's new Corvette—a car you could actually buy!

The 1954 Motorama gave no clue to the forthcoming 1955 Chevrolet, with one exception.

The Corvette Nomad combined the front end of the Corvette with the yet-to-be-released two-door 1955 Nomad. The styling cues, primarily the roofline and tailgate, were very much evident when the 1955 models were introduced in October 1954. After six years of "sameness" from Chevrolet, the 1955 models hit like a bombshell. So stunned was the buying public that initial response was rather cautious, and sales did not immediately take off. Soon, however, sales began to soar, and then there was no stopping the '55.

It marked a new era for Chevrolet and a popularity for the car that exists to this day.

CHAPTER ONE

The
Motoramic Chevrolet

Americans are often nostalgic about the 1950s, but the decade was not entirely one of peace. The Korean War had raged until October 1953, when a peace agreement was finally signed. America in the fifties wasn't only post-World War II, it was post-Korea, too. With the war finally at an end, Americans who longed for diversions could turn their attention from the headlines of the day to things domestic. And what was more domestic than cars? Among automotive diversions, few things could top General Motors' Motorama.

The Motoramas were the pinnacle of automotive glamour and excitement, with breath-taking show cars, eye-opening displays, and beautiful models waxing ecstatic over the new cars available at Buick, Cadillac, Chevrolet, Oldsmobile, and Pontiac dealers.

By 1955, the Motoramas were a country-wide event, traditionally opening in New York City. Opening day for the 1955 show was January 20. This was the 1955 models' first show appearance, and Chevrolet Division played it to the hilt. The 1955 Chevrolet was dubbed the "Motoramic Chevrolet" by the Division. This moniker was picked up by dealers around the country, and it appeared in Division advertising in such magazines as *Holiday*, *The Saturday Evening Post*, and *National Geographic*.

After its stint in New York, the Motorama caravan, made up of ninety-nine trucks hauling near-ly 1,200 crates along with 350 full-time employees, moved on to Boston, Miami, Los Angeles, and San Francisco. More than two million people attended these shows. It cost GM millions of dollars annually to put on the Motorama, but the company viewed it as money well spent. Undoubtedly they were correct, as nearly four million of the eight million cars sold in 1955 were GMs.

As usual, the big sales battle was between Ford and Chevrolet. In 1955, each claimed victory. Industry pundits charged a shell game was going on in just how the numbers were tabulated. Chevrolet had been the sales leader of the industry for nineteen straight years through 1954. Chevrolet ran an advertisement touting its sales success:

"Complete and official registration figures show that once again . . . for the nineteenth straight year . . . Chevrolet outsold all other cars! This great and growing record of sales leadership is a tribute to Chevrolet's unequaled popularity over the years, and to the accomplishments of the most alert and best qualified group of dealers in the automobile business—America's Chevrolet dealers!"

The 1955 Chevrolets were introduced in October 1954. In November, GM built its fifty millionth car. Because Chevrolet was GM's sales leader, it was chosen to take the honor.

The milestone, fifty-millionth car was a specially painted, gold 1955 Bel Air Sport Coupe. The

The popular two-tone paint scheme of the early fifties came into its glory with the 1955 Chevrolet. Bill and Barbara Jacobsen's 1955 Bel Air hardtop is finished in India Ivory and Gypsy Red.

The two-door Bel Air sedan and two-door Bel Air sport coupe were popular models within the Bel Air series. Nearly 169,000 two-door sport coupes were built in 1955.

To celebrate the production of its fifty-millionth car in 1955, General Motors pulled out all its publicity stops, including a parade through downtown Flint, Michigan. GM selected Chevrolet to carry this honor and painted the car, appropriately, gold. Thousands of people lined the parade route to watch the car pass. To see this photo is to understand America's love affair with the automobile in the fifties.

Low...and behold! A new concept of low-cost motoring!

This is what a new idea looks like.
This is Chevrolet and General Motors' entirely new approach to the design, engineering and production of a low-priced car!

Chevrolet

the motoramic Chevrolet for 1955

Turn the page and see the rest of the story; it's equally exciting!

celebration included a parade through Flint, Michigan, worthy of the achievement. Chevrolet ensured there was plenty of pre-parade hoopla. On November 23, 1954, on a crisp, clear morning, thousands of residents lined the parade route. First came a float with the twenty-five-millionth GM car, a silver 1940 Chevrolet Sport Sedan. Then came the 50 millionth car, the float surrounded by several dozen police officers to keep eager con-sumers at a safe distance. With a model waving from the driver's side, the parade watchers cheered and waved back.

The 1955 Chevrolet was off with a big bang!

The Hot One

The 1955 Chevrolet was the very fortunate re-cipient of a tremendous amount of brilliant engi-neering and design talent. The car truly was new

The 1955 Chevrolet was more than evolutionary—it was revolutionary. As the ads and brochures stated, the car truly was "new." Fortunately, the most enduring and pleasing styling exercise made it into production.

Edward N. Cole was not only a brilliant engineer, but a superb manager as well. He became chief engineer of Chevrolet in 1952. Cole assembled the engineering talent that designed the small-block V-8 that appeared in 1955, but this was just one area of his many engineering responsibilities for Chevrolet. In 1956, Cole became Chevrolet general manager

from the ground up. The styling was as right as it could be. If one looks at some of the early clay styling exercises, it becomes clear that the '55 could have been less than successful if certain designs had been implemented. The small-block V-8 was an economical yet powerful and compact powerplant that helped propel the car to many race wins and give the owners a measure of performance and prestige that added to the enjoyment of ownership.

How did it all begin? To start with, there was an industry trend in the early 1950s to V-8 engines, but Chevrolet lacked one. Chevys had been powered by a straight-six since 1929. The Chevrolet line had received all new styling in 1949, but by 1952, it was clear, this, too, would have to change. The "Transportation Unlimited" shows that evolved into Motorama were all the proof GM upper management needed to begin thinking about restyling the Chevrolet line-up. The designs for 1953 and 1954 were already committed by 1952, so it would be 1955 before the new line could make its debut.

But Chevrolet Division needed a change in thinking as well as a change in its product line.

Chevrolet was, in fact, working on a V-8 engine design under Edward H. Kelley's direction (Kelley had also worked on the design of the straight six). The V-8 project under Kelley was having its problems, and it was shaping up to be an expensive engine. This did not go unnoticed by GM President Charles E. Wilson and Executive Vice President Harlow H. Curtice.

Clearly, Chevrolet needed to be steered in a new direction, and the V-8 engine program needed better focus. GM upper management began their search, and one of the men considered to spearhead the new direction was Edward N. Cole.

A New Direction for Chevrolet

In the spring of 1952, Ed Cole was summoned to Detroit to meet with GM's upper management. Cole had the background for what they had in mind. Born and raised in Marne, Michigan, he joined the Cadillac Motor Car Division in 1929 as a co-op student at the age of twenty. He received a bachelor of science degree in mechanical engineering from the General Motors Institute in 1933. Cole continued with Cadillac, becoming its chief engineer by 1947 and directing the design of Cadillac's first V-8, introduced in 1949. He became works manager in 1950, and plant manager of the Cadillac-Cleveland tank plant that same year. In 1952, Cole received the phone call that changed his life and the direction of Chevrolet.

In discussions with the corporation's upper management at the General Motors headquarters

"Bel Air" script appeared on the right and left rear fenders above the spear point chrome trim, along with the die-cast Chevrolet medallion.

V-8-equipped Bel Airs carried this identifying piece of trim beneath each taillight.

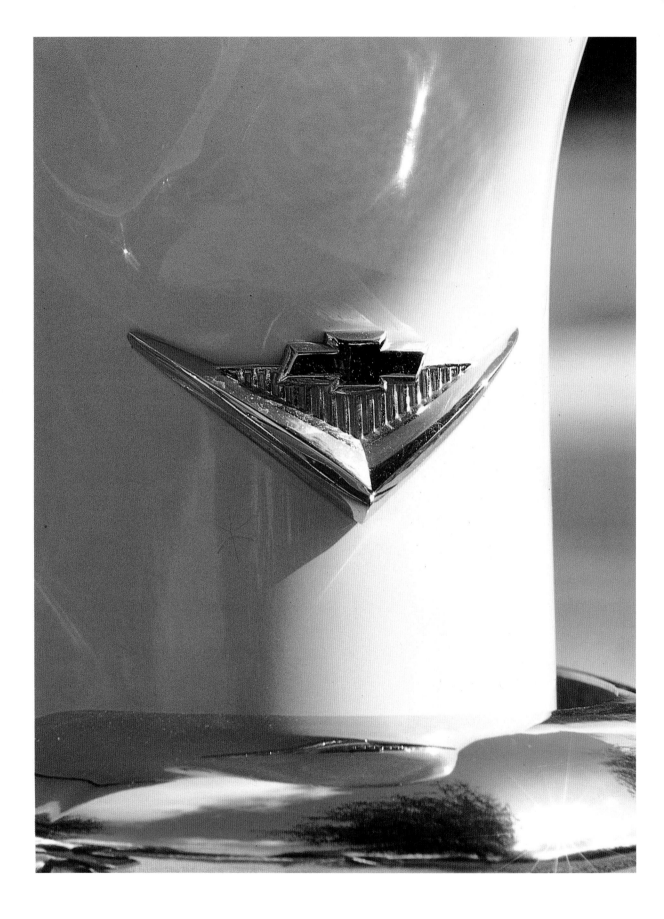

building in downtown Detroit, Cole proved he was exactly the kind of engineer for which they were looking. The position they offered Cole was manufacturing manager, but he felt he could better serve Chevrolet Division and achieve what was necessary as the Division's chief engineer. Edward Kelly had a heavy background in manufacturing, and Cole suggested that he move from his position as chief engineer to that of manager of manufacturing. Curtice and Wilson were impressed with Cole's engineering experience, savvy, vision, and drive. They were confident in giving the position of chief engineer to Cole, and Kelley agreed to become manufacturing manager.

Cole became chief of Chevrolet Engineering in May 1952. He had given considerable thought to staffing the Division to implement the countless engineering changes necessary for both the V-8 and the car itself, not to mention the affected trucks. Cole more than tripled the Chevrolet engineering staff of 850 to more than 2,500.

Harry F. Barr was one of the first people Cole hand-picked to come to Chevrolet. After receiving a bachelor of science degree from the University of Detroit in 1929, Barr had joined Cadillac Division. He became project engineer in charge of chassis and drivetrain in 1937. As staff engineer between 1946 and 1950, Barr helped Ed Cole develop the Cadillac V-8. With the outbreak of the Korean War, he was promoted to chief engineer of the Cadillac-Cleveland tank plant in 1950. Barr then moved on to become assistant chief engineer of Cadillac in 1952, a position he held when he was tapped by Cole to become assistant chief engineer of engine and chassis for Chevrolet.

Barr, in turn, selected Al Kolbe, already at Chevrolet Division, to be in charge of the new engine itself, aided by Don McPherson and other en-

Harry Barr (seated at desk) was assistant chief engineer working under Ed Cole. He became chief engineer of Chevrolet when Cole was promoted to general manager in 1956, the year this photo was taken. With him, from left to right, are Edward Gray, director of quality control; Russell Sanders, assistant chief engineer in charge of passenger car design; Ellis J. Premo, executive assistant chief engineer; Barr; and Byron H. Holmes, director of salaried personnel.

The Bel Air came standard with full
wheel covers, but the whitewall tires
were a factory option. This car is also
fitted with dealer-installed front
bumper guards and grille guard.

gineers. Kolbe went on to oversee the design of Chevrolet's first big-block V-8, the 348ci V-8 introduced in 1958. Don McPherson went on to promote and get approval for Mk IV 396ci V-8, which replaced the 348/409ci V-8.

The man in charge of body engineering for the 1955 Chevrolet was Ellis J. Premo. He had seen many changes at Chevrolet, having joined the Division in 1935. Premo had a wealth of experience in coordinating the efforts of the GM Styling Section, Fisher Body Division, and Chevrolet Engineering.

The all-important body designs for the 1955

car line fell to Clare MacKichan. He had received a mechanical engineering degree from the University of Michigan in 1938. His styling career began when he briefly studied in Harley Earl's styling school. MacKichan first worked for GM's Buick Division. This stint was interrupted by World War II, but he returned to Buick after a short period with Fisher Body Division after the war. He joined the Chevrolet styling studio working under Ed Glowacke. MacKichan became chief of the Chevrolet studio in April 1951 when Glowacke was moved to the Cadillac studio.

Working alongside these men, as well as those

The 265ci V-8 introduced in 1955 developed 162hp and with it The Hot One was born. The Plus-Power option, also known as the Power Pack option, bumped horsepower to 180 by using a four-barrel carburetor and dual exhaust.

The two-tone color scheme was carried over to the Bel Air's interior. The bench front seat was upholstered in cloth and vinyl, with the door panels all vinyl.

Next page
From front bumper to rear taillight, the Chevrolet styling team worked to produce a pleasing car. They did not know they were designing a car destined to become a collectable classic.

who reported to them, was an amazing array of talent, including engineers and stylists, not to mention those in manufacturing, making sure the cutting-edge technology was achievable.

Styling the 1955 Chevrolet

The styling of the 1955 Chevrolet car line was not solely the product of the Chevrolet styling studio or GM Styling, but a collaborative effort. The styling of the car took place in the GM research building on West Milwaukee Avenue in Detroit, behind the towering GM building headquarters.

Charles Stebbins was in charge of the 'A' body room of GM Styling, where Chevrolet and Pontiac body designs took shape under the ever-

watchful eyes of GM Styling Chief Harley Earl. The Chevrolet styling studio was on the 11th floor of the GM research building, with GM Styling on the 10th floor. Clare MacKichan was chief of interior and exterior styling for Chevrolet at the time work on the 1955 Chevrolet car line began in 1951.

Styling of cars had always been a subjective thing, a pure exercise in aesthetics dictated by the fashion of the times. Individual stylists within both GM Styling and the Chevrolet studio did countless renderings, with each rendering reflecting the stylist's vision of what the car could be. Nearly always, the renderings were done in the traditional, exaggerated technique that distorted the car's true dimensions for the sake of flair. This is true even to-

day. These renderings were flights of fancy but not totally divorced from production reality. They served as the styling basis that would evolve over the next two years before the purchase orders were made for the sheet metal stamping dies.

Numerous stylists contributed to the design of the 1955 Chevrolet, from the overall design of the bodies down to the smallest details. Among Tri-Chevy enthusiasts, the most familiar stylist is Carl Renner, whose official title during work on the 1955 car line was assistant chief designer. It is Renner's renderings that have most often appeared when initial styling exercises for the 1955 Chevrolet are printed. Renner's greatest claim to fame is his contribution to the design of the 1955 Nomad.

The styling of the 1955 Chevrolet was really an amalgam of individual ideas. Viewing photographs of some of the clay design bucks as the car evolved can leave one shaken from the possibilities of what might have gone into production. One photograph shows a version nearing completion, resembling very closely the final production design, except for the front end, which was heavy-handed and totally out of character with the rest of the car. Had this design gone into production, the 1955 Chevrolet would have been little more than a footnote in automotive history.

It is Harley Earl who is credited with the design of the egg-crate-type grille that graced the '55 front end, but it was this grille which was perceived as the problem when initial sales of the car were below expectations. Earl was vindicated when sales took off after two months of sluggishness.

The styling, or exterior design, of the 1955 Chevrolet had to work within the constraints imposed by Fisher Body Division and Chevrolet Engineering. Once a particular rendering was selected as a possible design, full-size drawings were generated, taking into account the aforementioned constraints, while reflecting the design goals of the renderings. Once this was done, full-size clay bucks were produced to evaluate the design in three dimensions. Much refinement and many changes were implemented at the clay stage.

Ed Cole was keenly interested in the design of the car as it evolved in Stebbins body room

Previous page
The gear selector above the steering column indicates this car comes with the two-speed Powerglide transmission. Note the decorative trim panel with the repeated Chevrolet bow tie pattern.

Chevrolet stylists mirrored the speedometer and dashboard details on the passenger side. Slightly different Bel Air script appeared on the black grille area which housed the radio speaker. The clock was a dealer-installed option, but proved difficult to read for the driver.

Next page
Another dealer-installed option was the tissue dispenser, costing a mere $3.85.

Art Imitates Life

From the very dawn of American automobile production, the medium used for advertising cars was not photography but illustration. Automotive advertising using color illustrations, or renderings, flourished during the fifties. While some photography was used in Chevrolet advertising, the artist's palette was used almost exclusively to display the new generation of Chevrolets ushered in with the 1955 model. These color illustrations were used both in advertising and in showroom brochures. Automotive illustration was literally an art form.

There were many pragmatic reasons for choosing illustration over photography. First, it gave the ad agency art director flexibility in selecting ad themes, allowing him to match the style of such magazines as *The Saturday Evening Post, National Geographic,* and *McClean's.* With an illustration, certain themes could be "staged" that would prove cost or time prohibitive for a photo shoot.

For example, the 1955 ad "Blue Ribbon Beauty that's stealing the thunder from the high-priced cars!" showcased a two-tone Bel Air convertible at an outdoor classic and sports car show. The car is receiving admiring glances from many of the well-to-do at the show, with numerous "high-priced" imported cars shown in the background. To find a suitable location, round up the required cars, arrange for the cars and their owners to be at a location on a given date, hire the models, and hope the weather would cooperate on the day of the photo shoot would prove costly and difficult to coordinate. If it rained that day, the photo shoot would have to be cancelled and rescheduled at additional cost. With an illustration, the only thing the account executive had to be concerned about was the illustrator himself. With sketches and an outline of the subject, the illustrator could produce the advertising art in a matter of days at reasonable cost—rain or shine!

Through illustration, the advertising director could juxtapose the car with other competing cars and put the cars and people in any locale desired. The car, most often a Bel Air, could be shown traveling along one of the new interstates, conveying the feeling of relaxing long-distance travel. Or perhaps a family was shown traveling across the rugged West with desert plateaus in the distance. Effortless hill-climbing was a theme used several time during the tri-Chevy era, such as the "The road isn't built that can make it breath hard" ad that appeared in 1957. The rolling hills depicted in the background could be almost anywhere, and the hill being conquered by the red, V-8-powered Bel Air pictured in the ad looks to be no less than a 20 percent grade.

All the ads that appeared during 1955, 1956, and 1957 showed middle-class Americans enjoying the good life. Despite the recession that put a damper on sales in the middle of the decade, America was in a postwar boom. Buying a new Chevrolet was just one way millions of Americans experienced that good life, and hitting the road for far-away destinations was another. Chevrolet made sure to embrace such life-changing events in a family as the husband racing to get his expectant wife to the hospital, taking a team of baseball players to a game in one of Chevrolet's spacious station wagons, or enjoying top-down driving with the wind in your hair.

Color illustrations also served as a means of showing the many two-tone color combinations available. This was advantageous when the showroom brochures or ads had to be produced prior to actual car production. How else to convey Glacier Blue with Shoreline Beige on a Two-Ten Townsman station wagon, Onyx Black with Crocus Yellow on a Nomad, or Adobe Beige with Sierra Gold on a Bel Air Sport Coupe?

In the fifties, automotive illustration shifted away from the photo-realistic air-brush renderings of the forties. The illustration style was more relaxed, and the media used included watercolor, pastels, and occasionally oils or acrylics. The cars shown were not as stylistically exaggerated as those rendered by chief designer of body development Charles Stebbins or done by assistant chief designer Carl Renner. The illustrations were contracted through Chevrolet's advertising agency and done by a freelance illustrator. Several of these illustrators made a name for themselves, as they were permitted to sign their artwork.

The illustrations themselves were simply tools used in advertising, and they were usually retained by the advertising agency. These days, any surviving examples are worth a handsome sum. These illustrations not only depict the cars themselves, but a bygone day shown in a rendering style that will probably never return.

The Bel Air convertible was the top of the Chevrolet line for 1955. To get that message across, Chevrolet's advertising agency produced this ad showing the car among some automotive blue bloods at an outdoor car show.

Chevrolet Bel Air Convertible with Body by Fisher.

Blue-ribbon beauty

that's stealing the thunder from the high-priced cars!

Wherever outstanding cars are judged a surprising thing is happening. The spotlight is focusing on the new Chevrolets!

Surprising—because Chevrolet offers one of America's lowest-priced lines of cars. But not really astonishing when you consider that its designers had just one goal—to shatter all previous ideas about what a low-priced car could be and do.

The unparalleled manufacturing efficiency of Chevrolet and General Motors provided the *means*—and that's why you have a low-priced car that looks like a custom creation. That's why you get the thistledown softness of Glide-Ride front suspension, the choice of a hyper-efficient 162-h.p. V8 engine or two brilliant new 6's. That's why Chevrolet's extra-cost options included every luxury you might want. And that's why you should try a Chevrolet for the biggest surprise of your motoring life! . . . Chevrolet Division of General Motors, Detroit 2, Mich.

Motoramic CHEVROLET *Drive it at your Chevrolet dealer's*

The chassis of the 1955 Bel Air convertible featured a massive X-brace to give the ragtop added rigidity. This had been a necessity for years to restore body strength integrity lost with removal of the hardtop.

and MacKichan's studio. It was vitally important to Cole that the car visually convey what he was trying to achieve in concept and engineering practice.

The main bodystyles the studio devoted itself to were a two-door sedan, four-door sedan, two-door station wagon, four-door station wagon, two-door sport coupe, two-door convertible, and two-door Nomad wagon, not to mention the utility vehicle variations and trucks that did not appear in the brochure for the cars.

The trim levels included the base One-Fifty series, the higher trim level Two-Ten series, and the top of the line Bel Air, which included the convertible and Nomad. Integral with the trim levels were an array of interior appointments and color combinations.

The ride, handling, and structural integrity of these various models was yet another aspect of the vehicle program, and this was an area that Ed Cole and Harry Barr were determined to improve over the 1954 and earlier models.

Chassis and Body Engineering

Beneath the handsome sheet metal of the new 1955 models was a completely re-engineered car. This would not be evident to buyers, and, in fact, might hold only passing interest to them, but it was vital to the car's operation and handling. Still, these changes and improvements were significant, and many made their way into the showroom brochure and were used as selling points.

Harry Barr oversaw the engineering of the chassis and suspension. Cole and Barr decided to make a break with the previously used K-type frame, which was made up of two channels for the main frame and K-sections. The engineers wanted a lower rocker sill height. To achieve this, they envisioned putting the frame, or chassis, within the rocker sill instead of beneath it, effectively lowering it in relation to the ground.

Chevrolet's heavy manufacturing facility in Flint worked closely with the engineering group to develop an ingenious way of achieving this goal while strengthening the chassis. Large-diameter steel tube was formed into a rectangular section by presses and then bent and formed to provide clearances and proper location for attaching other components and welded parts. The exact configuration of this chassis and its relation to the body had been worked out using full-size wood mockups. The convertible received a massive X-brace to compensate for the lost roof structure of the hardtop. Initial calculations predicted a stronger chassis, more resistant to bending and twisting.

The first frame was delivered to the engineering lab in February 1954. Testing confirmed it was 18 percent lighter and 50 percent more rigid than the chassis of the 1954 car, a stunning engineering achievement.

The live rear axle suspension remained relatively unchanged, but the front suspension was new. It featured anti-dive geometry and used ball joints in the steering linkage.

Ellis J. Premo was assistant chief engineer for the 1955 body. Premo had years of experience in this area at Chevrolet, acting as liaison among Chevrolet Division, Fisher Body Division, and Styling to ensure form followed function on the new car.

For the 1955 car, all panels and pillars that surrounded the door openings were framed as a single, rigid unit for assembly to the body shell. This unitized construction meant better body alignment and better fitting doors. In the two-door and four-door sedans, a heavy-gauge steel floor brace was welded between the center pillars, while a roof bow tied them together at the top. The entire body structure, from the firewall back, was bolted to the chassis with rubber body mounts to reduce harshness and quiet the ride.

In the sales brochure for the 1955 car line, Chevrolet touted this engineering feature as a strong sales tool: "There's a lot more that's new besides the beauty in Chevrolet's sleek new Bodies by Fisher! The entire body structure has been redesigned and more closely integrated with the frame. Together with new manufacturing methods, this new design brings you . . . new strength, rigidity and roominess. And it contributes greatly to the solid and steady big-car comfort you'll enjoy in the new Chevrolet. Remember . . . Chevrolet's the only low-priced car to bring you the advances and advantages of the body offered in many of America's highest-priced cars—Body by Fisher!"

It took an exhaustive test and development program at the Milford Proving Grounds and countless drives around the country in test mules to work the bugs out of the new body and chassis combination.

Rosenberger remembers just how arduous it was: "The new engine and body mounts in the car and the silencing of the powerplant took a tremendous amount of work," because all the attached accessories had a proclivity for vibrating at some speed that's resonant with something in the car.

The 1955 Chevrolet's ride and handling were only as good as its chassis and suspension. Cole oversaw the re-engineering of the Chevrolet chassis so it could accommodate both the venerable inline six-cylinder as well as the small-block V-8.

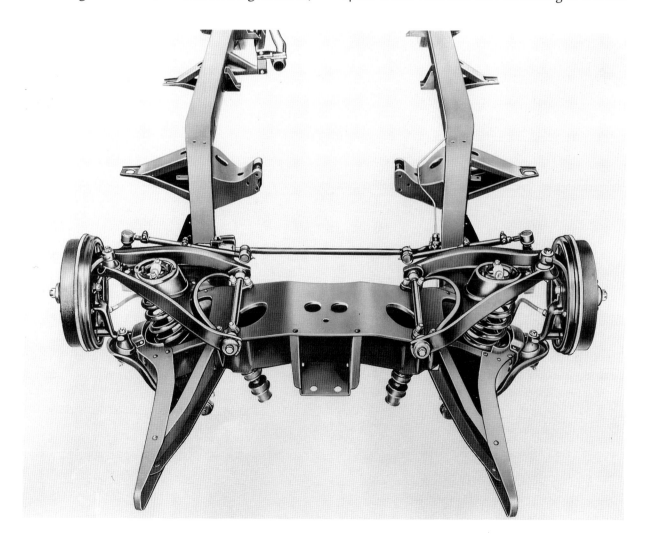

The Bel Air convertible was the glamour girl of the 1955 car line. This particular example, owned by Dennis and Mary Schrader, is a rarity because it was not ordered in a two-tone color combination.

The stylized, winged hood ornament on the 1955 Chevrolet was designed by Carl Renner, assistant chief designer working under Clare MacKichan, who was chief of the Chevrolet design studio.

You have booms and buzzes and things that drive you nuts. A guy like Cole, coming from Cadillac, wasn't putting up with that stuff at Chevrolet. We worked our tails off on things the average guy, building the lowest car on the rung of the ladder, probably wouldn't have worried about. I can remember we were working weekends and nights on things that annoyed us. You drive up a 7 percent hill, and if you happened to drive a certain speed, it would boom so bad it would drive you nuts. We had dozens of things like that to work on. I think it helped our reputation to put out a slick car right from the start compared to what we would have done if we hadn't had a guy like Cole around."

The Small-Block V-8

Central to the 1955 product line was the V-8 engine to be designed for the new car. As mentioned earlier, Ed Kelley had been working on a V-8 design, but when Ed Cole studied the engine, it was clear it had to be abandoned. Kelley's engine was

This cross-sectional drawing shows the oiling system developed for the small-block V-8. This design feature was a vital aspect of the engine's reliability and longevity. The hydraulic lifters pumped oil up through the pushrods, past a hole in the rocker arm, and over the valve train components. The stamped-steel "umbrella" over the valve spring was conceived during development to eliminate oil burning.

patterned after the 1949 Cadillac and Oldsmobile V-8s; it featured a heavy design with thick casting walls and employed an expensive rocker shaft valvetrain. Ed Cole had the opportunity to oversee the design of a completely new, lightweight V-8 engine, and he relished the challenge.

The design of the small-block V-8 was started

at the Chevrolet Central Office (a four-story annex behind the landmark General Motors Building in downtown Detroit) in the ground-floor drafting room. Fred Frincke joined Chevrolet in 1946 as a drafting trainee and remained with the Division for forty-four years. He worked on a number of Chevrolet's most famous engines and has many

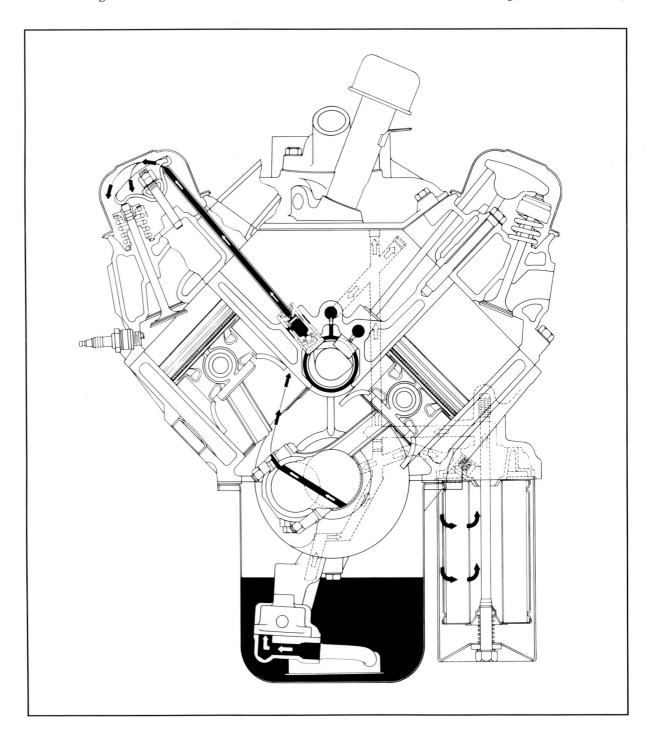

Key to the 1955 Chevrolet's success was the compact but powerful small-block V-8. The 265ci V-8 was a marvel of simplicity and economy. Enthusiasts saw the performance potential of the new V-8, and Chevrolet itself quickly joined the fray by boosting the small-block's output each year. The engine became a focal point in Chevrolet's advertising, offering the most bang for the buck while being fuel efficient.

Mk IV big-block engine developments to his credit.

"I remember the day Ed Cole came to the drafting room after coming to Chevrolet," Frincke recalled. "He introduced himself to everyone in the drafting room and said, 'Instead of Chevrolet engineers driving Cadillacs, Cadillac engineers are going to be driving Chevrolets.' I remember that verbatim. Back at that time, Chevrolet engineers were buying Cadillacs because, under the GM discount program, you could get a big discount on a Cadillac."

It wasn't clear whether Cole meant that engineers at Cadillac would soon opt for driving the V-8 Chevrolet slated to appear in 1955, or whether he meant he would pull many of the Cadillac engineers he had worked with over to Chevrolet—but it quickly appeared the latter would be the case.

The drafting room at Chevrolet became a hotbed of activity after Cole arrived. One of those in the thick of it all was Dave Martens.

"I started as a college graduate student-in-training in January 1952," said Martens. "For the next two years, I was alternately a draftsman and worked in the various plants and laboratories, then I was a draftsman full time. At the time Ed Cole and Harry Barr came to Chevrolet from Cadillac, Chevrolet had, as I remember, a 231-cubic-inch V-8 engine on the boards, and they even built a couple. When Cole came in, they scrapped that design and started over, more or less on the basis of how the Cadillac 331-cubic-inch V-8 was done, which he had worked on.

"Al Kolbe was the assistant staff engineer in charge of the design of the 265-cubic-inch V-8. He was low-key—a pure designer—and would spend most of his time in the drafting room, sitting down at the drafting boards with the designer, suggesting this, suggesting that, sketching what he wanted. The configuration of the 265-cubic-inch V-8, I would have to say, was basically [made up of] his ideas. Yes, Ed Cole was responsible for ramrodding the program through, but Al Kolbe did the actual design work. He did not want to get any notoriety. He wasn't interested in writing an SAE paper or any of that—he was a pure designer. Al Kolbe was, in my opinion, not properly recognized for the design genius that he was," Martens says emphatically.

"Kolbe would be in the drafting room every day," Martens remembered. Cole would visit the drafting room once a week because his activities were running a lot more at Chevrolet than just the new engine design."

As chief of passenger-car chassis design for Chevrolet Motor Division, Russell F. Sanders worked closely with the engine design engineers from the very beginning. He received his bachelor of science degree in mechanical engineering from Michigan State College in 1929, then joined the Oldsmobile Division of General Motors. In 1934, he was assigned to the Chevrolet Motor Division. When it came time for Chevrolet to begin its work on a V-8 engine, Sanders met continually with the engine team to ensure engine-to-chassis compatibility. He wrote a paper for the SAE, chronicling the development work on the Chevrolet V-8. From the outset, this engine was conceived and designed with different goals in mind from those that were the norm in V-8 engine design at the time.

Many discussions centered on displacement because it established bore, stroke, and overall block dimensions.

"In our research over a period of years," Sanders wrote, "we have investigated many types of V-8 engines. During the early stages, we developed an engine with 231 cubic inches of displacement, but with changing conditions, a greater displacement was considered desirable.

"At one of our group meetings, we sketched some basic outlines to indicate just what we wanted in height, displacement, length, and so forth. We began thinking about a 245-cubic-inch engine, but when we got further into this study we found we could just as well go to 260 or 265 with no penalty of extra weight, knowing that ample displacement is fundamentally the most economical

Previous page
The exhaust extensions were an inexpensive dress-up option available through the dealer.

The basic dimensions of the small-block V-8 have remained essentially unchanged since 1955. It became one of the most successful and prolific automotive V-8 engines of all time, with tens of millions built.

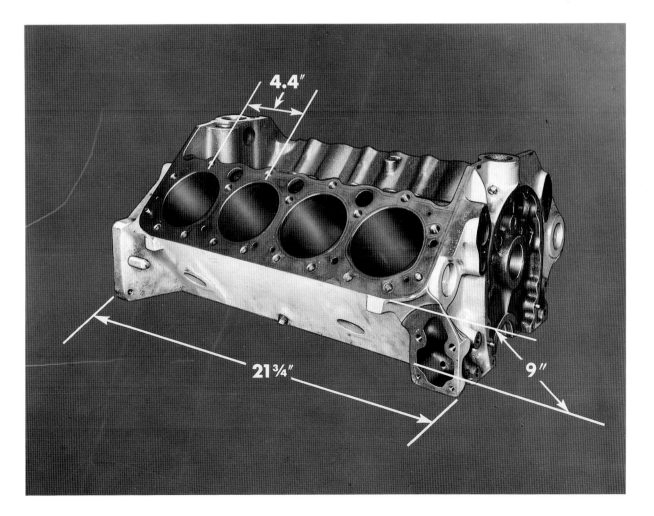

4.4"

21¾"

9"

The small-block V-8 engine block began life as an assemblage of green sand cores which were placed in jigs and lowered into the drag mold prior to pouring the molten iron. This aspect of engine casting has changed little over the decades, though it is done with much more precision today.

way to insure high torque and resultant good performance economy."

After much discussion, a bore size of 3.75in was selected. To achieve a 260ci displacement, a 2.93in stroke was required. The engineers decided to round this off to 3.00in, which resulted in a 265ci displacement for the engine. Cylinder bore centers were fixed at 4.4in.

With this information, the length of the crankshaft was determined. After factoring in the width for main bearings, connecting rod bearings, crankshaft arms and engine block bulkheads, the overall length of the engine block was 21.75in.

Kolbe felt, and Cole agreed, it was unnecessary to carry the block more than 0.125in below the center line of the crankshaft. This was just

enough, in their view, for solid placement of the main bearing caps in a machined notch and was in line with their goal of light weight.

Chevrolet refined its V-8 casting procedures with its prototypes. The 265ci V-8 block was meant to be a precision casting. Reducing the number of cores reduced the tolerance error resulting from core stackup variations.

"Fewer cores," Sanders wrote, "means that our section thicknesses can be controlled much more accurately, and we have less sand to handle. The end result is a precision casting which is lighter, as well as lower in cost."

John Dolza of GM's engineering staff is credited with the internal structure of the engine to reduce the number of sand cores of the new Chevrolet V-8. He patented a process of casting called the "green sand" process in which damp sand is repeatedly tamped around a solid pattern to produce the sand core required for casting. This differed from the traditional process of mixing sand with a binder and letting it cure, or bake, in its separable mold to produce the core needed for casting. "Green sand" casting was a less-involved process, was more economical, and permitted fewer molds to produce a cast-iron part.

There were nine major and three minor cores used to cast the 265ci V-8 block. A competing manufacturer required twenty-two cores to cast its V-8. Dolza reduced the number of cores required by mating the left, and right-hand cylinder-barrel cores in a "V" with an integral crank chamber core. These four cores were put on a slab. Two, one-piece jacket cores were slipped over each bank of cylinder cores. The jacket cores were supported in position and prevented from moving by the end cores of the block. These end cores, also mounted on the slab, formed the contours of the timing chain case and

This photo taken at the gray-iron foundry shows iron being poured into drag molds with the engine block sand cores. After proper cooling, the sand cores were broken and removed, and the raw castings were sent to the machining line.

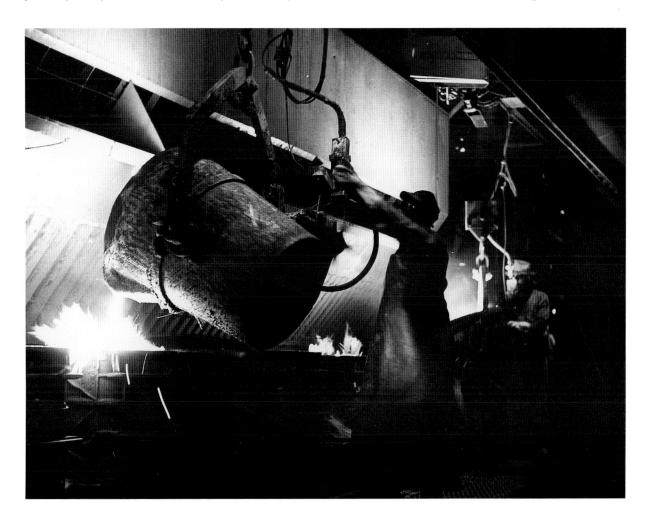

Each cylinder bore in the machined, bare blocks was checked by this Sheffield air gauge. Results were indicated on the machine directly in front of the operator. The engine blocks then proceeded to the engine assembly phase.

front structure, and at the rear, the clutch housing attachment configuration.

The entire assembly of cores was placed in a drying fixture. After drying, it was moved to the casting line, known as a drag mold. Any loose sand was blown out and pouring of molten iron was performed in a smooth operation. After cooling, the cores were broken and removed, and loose sand dislodged. The casting then moved to the machining phase. The fully machined 265ci V-8 block weighed only 147lb, compared to 163lb for Chevrolet's in-line six-cylinder engine.

The Cylinder Head

With an eye toward cost and proven performance characteristics, engineering chose the wedge-type combustion chamber with in-line valves for its cylinder head.

"We settled on the high-turbulence, wedge-

type design for combustion control and combustion smoothness," Sanders wrote, "since it controls the rate of pressure rise in the chamber. We feel that the type of chamber which exposes a high volume of the charge early in the burn cycle and then goes out into a quench area gives low octane requirement and smoothness of operation."

The designer of the cylinder head for the 265ci V-8 was Don McPherson. He came to Chevrolet in 1940, and alternated between working for Chevrolet and working toward a bachelor's and then a master's degree, with time off for a service stint during World War II. He went to work for Chevrolet Engineering full time in 1948. In 1950, McPherson was a drafting layout man in the engine group. Work was begun on the V-8 design under Ed Kelley and in 1952 he became drafting supervisor. He saw Cole and Barr on a regular basis.

"Cole came in one day," McPherson remem-

bered, "and said, 'I want the smallest, lightest-weight cylinder head you can come up with.' This was back when I was design supervisor, but they put me back on the board to do that head, about 1953. I made sketches of the cross-section of the head—a whole bunch of them—and the smaller I made them, the better Cole liked them. Finally, I had a cross-section of the head where we just barely had room for the rocker arms and the rocker arm cover, and he bought it.

"On the cylinder head itself, a big problem there was squeezing in enough water, by our standards, around the valve guides," McPherson explained. "We ended up, actually, with less water around the valve guides than any engine at the time, and that was more a tribute to what the foundry people were able to do, I think, than what we were able to draw. That was a real, real tight job.

"The biggest problem with the cylinder head," McPherson continues, "was that Cole wanted it very small, and he didn't want excess machining on there. He wanted a cast surface on the top for the rocker cover gasketing surface. That was the first time that had ever been done. Except for the spring seats, there was no machined surface on the top because the bosses for the rocker arm studs were higher than the basic surface for the rocker cover, so there was no way, without getting a traversing mill that went around there, to do that thing properly. We ended up with just a cast surface there, and then we had to develop gaskets for that dog-

Next page
Despite Chevrolet's advertising claim that it was the "leader in the low-priced field," this dressed-out Bel Air convertible turned heads wherever it went—and still does today. The car's lines still please the eye forty years later.

The engine compartment of the Jacobsens' show-winning Bel Air convertible gives admirers a perfect picture of what was under the hood of the Chevrolet in 1955. It is authentic down to the last hose, gasket, and seal.

gone thing, and we never did a particularly good job of it.

"Another problem we had with the cylinder head," McPherson added, "was how do you get the oil back down into the crankcase? We designed a trough at the back end that would drain down through the block, but the blow-by of the engine was so high that the blow-by would keep the oil from coming back down the engine. This was a very severe problem. What we ended up doing was casting some chambers in the back of the cylinder head, then we drilled holes down through the cylinder head into the crankcase so we didn't have all that blow-by pushing oil back up. That was a last-minute change that we rushed into production."

Each cylinder head used a total of seventeen bolts to secure the head to the block. A five-bolt, pentagon pattern was used around each combustion chamber, instead of the usual four bolts.

When all the sand cores of each cylinder head were assembled, the fixture was inverted before being put into the drag mold. The cylinder head was cast in this manner to get harder iron at the "bottom" of the head, around the valve seats. The combustion chamber was cast to shape; the only machining necessary in the chamber was for the valve seats.

Rocker covers are not often an aesthetic concern, but on Chevrolet's new V-8, Cole felt they should be special, within cost parameters. It's a

The automobile restorers' craft is amply displayed inside and out on the Jacobsens' '55 Bel Air convertible. Their car is certainly one of the finest examples in the United States today.

Even the horn ring received careful design attention in the Chevrolet studio. When the V-8 was installed in your Chevrolet, an emblem was affixed to the steering wheel hub.

funny story, one McPherson retold with relish.

"Cole came in one day and said, 'I want a sexy rocker cover.'" It should be stated here that McPherson was noted at Chevrolet for his dry sense of humor, and could be relied upon to respond with a perfectly straight face that would leave other engineers aghast. He answered Cole by suggesting the ends of the rocker covers resemble the upper front superstructure of the female anatomy. Cole was not amused. What McPherson came up with was a rocker with handsome Chevrolet script across the top."

The AM pushbutton radio with antenna was a dealer-installed option costing $84.50.

Next page
The 1955 Nomad was inspired by the Corvette-based Nomad show car, first displayed at the 1954 Motorama. The roofline of the production Nomad was lifted almost line for line from the show car. Note the chrome trim above the headlights, exclusive to the Nomad. The spotlights were usually ordered only for police cars. The exterior windshield visor was a dealer-installed option.

The optional wire wheel covers, available on all Chevrolet passenger cars, gives this Nomad added sports car flair.

The exhaust manifolds were located near the top of the cylinder head, with the manifold going up, over, and then down in the center, a pattern with which Chevrolet enthusiasts of the fifties are familiar. This made for very short exhaust passages in the cylinder head, and with less cast-iron, the cylinder head retained less heat.

Intake Manifold

Chevrolet engineers studied the intake manifold to combine a number of functions and try to clean up the appearance of the top of the engine. Besides its fuel intake function, the Chevrolet intake manifold was designed in such a way as to seal the top of the engine, eliminating the need for a separate tappet-chamber cover. Instead of open-

ings between the intake passages—as was typical with other makes—there were webs. This one-piece design also eliminated complex parting lines during casting, but the resulting "flash" required additional grinding.

Reciprocating Parts

Chevrolet studied both the alloy-iron, or cast, crankshaft and the forged crankshaft for its new V-8. Forged steel was selected because of its high modulus, or degree, of elasticity and its specific gravity. Chevrolet could also employ its vast forging capacity at its facilities. New forging processes were used to reduce draft angles and make the crankshaft as short as possible.

"Our crankshaft design made it possible to produce what we believe to be the smallest and lightest commercial V-8 power package for its displacement," Sanders stated in the SAE paper.

The flat-top piston was aluminum with a piston offset toward the major thrust face to avoid piston slap and give ample tolerances on piston fit for manufacturing economy.

The piston was a three-ring design. Chevrolet spent a considerable amount of time examining the oil consumption problem under high vacuum conditions such as deceleration. They found oil passed the piston rings not between the ring and the cylinder wall, but between the ring and the piston ring groove. Chevrolet solved this problem, af-

The seven chrome trim bars mounted on the tailgate were also featured on the Nomad show car. The vertical handle in the center of the trim bar opened the tailgate. The license plate says it all.

ter numerous different designs, with a new type of expander between two oil control rings in that piston ring groove.

The connecting rod had a 5.7in center-to-center distance. The piston pin was pressed into the connecting rod during assembly. The connecting rod underwent eighteen million cycles in testing rather than the usual ten million with 2,400lb tension and 7,000lb compression without a failure.

Valvetrain

Rather than go with rocker arms mounted on a single rocker shaft, Chevrolet took a different approach. The designer of the Chevrolet swivel-ball rocker arm system was Clayton Leach, who later became assistant chief engineer of Pontiac. Each rocker arm was a precision stamping which pivoted on a stud pressed into the cylinder head.

The rocker arms were made from 1010 steel, carbon-nitrited and hardened to a depth of from 0.01 to 0.02in, then surface-treated. Once assembled over the valve stem and pushrod, the rocker arm was retained on the stud with a fulcrum ball

and locknut. This made it supremely easy to adjust each valve for lash, regardless of whether the lifters were hydraulic, as was the case here, or mechanical, which came on V-8s with manual transmissions.

Proper oiling of the valvetrain was something Cole wanted to ensure, particularly with the use of the stamped-steel rocker arms.

"The big development on that thing was how to get oil up to the rocker arm," McPherson said. "The fellow who worked on that was Bob Papenguth. He did the work on the wafer in the tappet that metered the oil up the pushrod to the rocker arm. That was the key to the success of the thing. In the past, they had always had complicated systems that got the oil topside and down through the center shaft on the rocker arm. But this thing took the oil up through the tappet and the pushrod and around the ball. Without that development, it would have been an impossibility."

V-8 Research and Development

Work on the 265ci V-8 proceeded quickly and by the spring of 1953, prototypes were running in

Chevrolet's engineering lab on Holbrook Avenue in Hamtramck. Dick Keinath was right in the middle of his two-year training program when he became involved with the 265ci V-8's valvetrain testing. Keinath performed the first deflection studies on the rocker arms.

"We were doing deflection work to see how strong the studs, stud attachments, the rocker arms, and the pushrods were," he remembered. Keinath was yet another new engineer in the right place at the right time at Chevrolet, which permitted him to become involved in one of the most exciting engineering programs in the company's history. Because he was fortunate enough to work on

Chevrolet's new V-8, he eventually went on to become assistant staff engineer, then chief engineer of engines years later.

In 1950, however, Keinath was just a graduate of Michigan State University with a bachelor degree in mechanical engineering. Like many others who applied to General Motors, he had worked on and had rebuilt automobile engines as a teenager. That year, he began working in General Motors Product Study in the area of automatic transmissions. These included the Buick Dynaflow and Chevrolet's Power Glide.

In 1952, Keinath transferred from GM's engineering staff to Chevrolet and began that Division's

The small-block V-8 was the obvious choice to power the Nomad. This rare car has the seldom-selected air conditioning option which was first offered in 1955. Air conditioning cost more than $500 at the time, so few Chevrolet buyers checked it off.

The original owner of this Nomad spared no expense to add to his driving pleasure. Power steering, option code no. 324, was a $91.50 option.

Next page
Chevrolet ran numerous ads for 1955 with the theme, "Stealing the thunder from the high-priced cars!" The introduction of the 265ci small-block V-8 gave the copywriters that much more to write about. This ad highlights the three basic engines available in the 1955 Chevrolet: the Turbo-Fire V-8, the Blue-Flame 136, and the Blue-Flame 123. The car featured in the ad was the Bel Air four-door sedan.

very extensive training program. The program exposed the engineer to all phases of automotive design engineering, manufacturing, and testing.

"When I transferred to Chevrolet," he explained, "I joined the training program. The training period included a three-month assignment at the design center, a three-month assignment in the laboratory, where I did work on the small-block V-8, a three-month assignment at the GM Proving Grounds, three months at the engine plant in Flint, Michigan, a period of time at the gray-iron foundry in Saginaw, Michigan, another couple of months at their manual transmission plant in Saginaw, another three months at the automatic transmission plant in Cleveland, Ohio, some time at the Tonawanda, New York, engine plant, some time at the gear and axle plant in Detroit, and three months at the metallurgical laboratory in Detroit. Then it included another period of time actually in the V-8 engine design group in Detroit across the street from the GM building in what we lovingly referred to as 'The Bank Building'.

"In 1954, management took the engine design group and the transmission design group and removed them from the main General Motors building and rented this old bank building across the street which was about 100 years old. They had old tapestries on the walls, a ceiling about forty feet high, columns in the front—an old, old building. It was dirty, drafty, the most unlikely building to use as a design center, because everyone had to have their own fluorescent lamp to get light on their own board. The bathrooms didn't work, and when there was a heavy rain, the basement flooded. When the front door opened, the breeze would come in and blow dirt, dust, and everything all over the boards. Anything you happened to have on the boards would fly away."

Although ground had been broken for a new Chevrolet Engineering Center in Warren, Michigan, north of Twelve Mile Road, engineers and designers would have to endure the old building for two years before the complex would be open.

Development testing of the new V-8 was well underway when Fred Sherman joined Chevro-

THE BEL AIR 4-DOOR SEDAN

Chevrolet's 3 new engines put new fun under your foot!

You've got the greatest choice going in the Motoramic Chevrolet! Would you like to boss the new "Turbo-Fire V8" around . . . strictly in charge when the light flashes green . . . calm and confident when the road snakes up a steep grade? (Easy does it—you're handling 162 "horses" with an 8 to 1 compression ratio!) Or would you prefer the equally thrilling performance of one of the two new 6's? There's the new "Blue-Flame 136" teamed with the extra-cost option of a smoother Powerglide. And the new "Blue-Flame 123" with either the new standard transmission or the extra-cost option of new Touch-Down Overdrive. See why Chevrolet is stealing the thunder from the high-priced cars? It has that high-priced, high-fashion look and everything good that goes with it—power, drives, ride, handling ease, everything. Let your Chevrolet dealer demonstrate how Chevrolet and General Motors have started a whole new age of low-cost motoring! . . . Chevrolet Division of General Motors, Detroit 2, Michigan.

Stealing the thunder from the high-priced cars! **Motoramic** CHEVROLET

let in April 1954. He recalled what the old dyno facilities were like: "I started in the dynamometer cells. "The V-8 was underway as far as development, almost a preproduction version of what was going to be released in the 1955 Chevrolet. The engines were measured on the dynamometer with open exhaust, which were tubes that were 12 feet long and about 2-1/2 inches in diameter, spark-adjusted and mixture-adjusted for maximum torque, corrected for 60 degrees Fahrenheit. Fan horsepower was added to that, and that produced a figure that was at that time called advertised, or gross, horsepower."

"We were at Holbrook and St. Aubin, which is now part of the gear and axle complex. At the old complex, we had exposed control panels with

Previous page
Before the Chevrolet Engineering Center opened in the mid-fifties, the Division had struggled with outmoded facilities and old equipment. The new center offered vastly improved facilities for the research, design, and engineering of new engines and vehicles. The front portion of the building held the engineering offices. Visible along the entire south side of the building are the engine dynamometer test cells. Behind the dyno cells are the machine shops and garages.

Power hydraulic brakes greatly reduced the effort required to slow this Nomad. First-time power brake drivers often brought the car to a screeching halt.

Chevrolet made sure the Nomad would not be confused with the two-door Two-Ten Handyman wagon by adopting the rakish rooflines of the Nomad show car and enlarging the rear wheel openings. This pristine example, painted India Ivory and Skyline Blue, is owned by Erol and Susan Tuzcu.

all the contacts for the rheostats. The dynamometers there would no longer handle output of the engines we were testing. At one point during the testing of a Corvette engine, the dynamometer actually overheated and began throwing flames out the back! They were fine for the straight sixes, and I imagine that lab had been there when they still had four-cylinder Chevrolets."

During 1954, many 265ci V-8-equipped Chevrolets were driven extensively to expose any

engine or vehicle problems. There were numerous vexing problems which Ed Cole and Harry Barr would not tolerate and these had to be worked out. These problems often surfaced during the rigorous test drives.

Maurice "Rosey" Rosenberger, chief experimental engineer and one of GM's most colorful and dynamic employees, actively participated in these test drives. His perspective on the V-type engines of the corporation is unique because it dates

The Nomad had a unique selection of two-tone vinyl seat interiors. The Tuzcus' Nomad is shown with Ivory and waffle-pattern Turquoise vinyl.

back to the 1920s. His is a fascinating story, and it isn't often one can listen to a man of his experience recall the original development of General Motors' V-8 engines: "I worked in a Ford assembly plant in Omaha, Nebraska, for a while in 1926." "I was probably as experienced as the average farm boy, being raised on a farm around farm equipment. Keeping an old Model T running to drive to school was a big deal," he laughed.

Rosenberger decided that if he was going to work around automobiles and engines, he may as well do it in the automotive capital of the country—Detroit. He left the rich farmland of Nebraska and struck off for Michigan in the fall of 1926.

"I joined Cadillac in 1927 as a mechanic on the engine assembly line," he said, picking up the story. "They were building the first V-8 LaSalle engines. They had built about 300 or 400 of those and decided they wanted to change rings and change bearings and a few other things, so they set them aside. They needed extra help to rebuild them. I was hired in as one of the mechanics to work on those things. I was twenty-four years old.

"When those engines were rebuilt, I had a chance to go to the experimental department on Clark Avenue in Detroit, and that was my first experience dynamometer testing engines. When I hired in at Cadillac, I got to work on better equipment than I ever had before. The V-16 was being tested in our lab in 1928 and 1929. The V-12 came into being some months after the V-16 was well into our test program. The Cadillac staff began building a completely new engineering, design, and testing building on the east side of Clark Avenue in 1931.

During the thirties, Rosenberger was involved with transmission design. Automaking was shut down in February 1942 to concentrate on war materiel production.

During the war years, Rosenberger worked on a program to adapt two Cadillac V-8 engines to a special transfer case transmission to power light tanks contracted by the War Department. As quickly as Cadillac automobile production had stopped in 1942, it started again with the end of the war, for the 1946 model year. He continued to work for Cadillac for the next six years. Rosen-

The Nomad did not sacrifice utility for the sake of sportiness. It's carrying capacity matched that of the Handyman wagon. With the rear seat folded down, the cargo area swallowed everything from surfboards to ladders. The spare tire is stored beneath the hinged compartment near the tailgate.

berger had distinguished himself at Cadillac, and this hadn't gone unnoticed by Ed Cole. In 1952, Rosenberger moved to Chevrolet and started work in the experimental department. There, he worked on isolating the troublesome noises the 265ci V-8 generated in the prototype cars, as well as on general car development.

"Cole was organizing trips," Rosenberger remembered, "and would go on trips with us, for a day or two. I remember going down through Ohio one day in 1954, I think it was. We had a bunch of cars and started on a trip. We got out a ways and many of them were smoking so bad, he said, 'Let's turn around and fix this damn thing.' The smoking was due to ring design, and some oil coming down the valve stems. We went back and probably put in a month or two of work and took another trip."

Sherman remembered not only the oil burning problem the engines were experiencing, but Cole's active participation in solving it: "At the old lab, I remember Ed Cole coming in one time when we had this engine oiling problem. He came in there while we were working on one of the engines. He had his suit coat off, his sleeves rolled up, and he was working right on the engine with us. Of course, we all called him Mr. Cole, which he didn't like, and he would say, 'Just call me Ed.' We did, but we were just a little hesitant about it. Finally, we did come up with a fix for the oil burning in that we put a sheet metal umbrella over the valve spring to shield the oil splash from going down the intake valve and the valve guide."

Concurrent with the V-8 engine design and development was the program for the car's body styling and interior design for all the various models, and chassis development.

New Look! New Life! New Everything!

There's no doubt Ed Cole's stamp was visible throughout the new Chevrolet, but he was also blessed to have an astounding engineering and styling team working with him. All these elements came together in the 1955 Chevrolet, helping to make it one of the best-selling cars in its day and enduring in popularity today.

Chevrolet dealers all across the country had been eagerly anticipating the arrival of the new

1955 car line. Their appetite had been whetted when several thousand dealers were flown to Detroit in July 1954 to see the cars they would soon be selling. The cars had been the subject of much magazine editorializing and coverage. The arrival of the first trailers with the 1955 models at each dealer was met with a level of excitement that had been absent for years. May school boys ran home bursting with unbridled excitement after spotting the new '55s—"Mom! Dad! The new Chevys are in!" And there were 7,500 Chevrolet dealers across America where this excitement was repeated in the months after the line-up's October 1954 introduction. Suddenly, the 1954 models looked ancient.

At the New York opening of the 1955 Motorama in January of that year, the 1955 Chevrolet was dubbed "the Motoramic Chevrolet." That theme was repeated in Chevrolet advertising. The 1955 models were a public relations and advertising agent's dream. And Chevrolet didn't skimp on advertising! Ads for the car appeared in countless magazines, from *National Geographic*, to *Maclean's Magazine* and *Holiday*, to name just a few.

Previous page
These handsome trim bars on the headliner echoed the subtle ribs in the roof of the Nomad. This, too, was borrowed from the Nomad show car.

There were fourteen Chevrolet models to choose from in 1955. The four-door Bel Air was the most popular of them all, selling more than 366,000 units.

A simple foldout-type showroom brochure stated "New Look! New Life (V-8 or 6)! New Everything!" followed by a brief description of each model and the engineering features. A much more detailed showroom brochure described all the models, showing the different trim levels, color combinations, engineering features, and specifications.

There were three distinct series Chevrolet offered for 1955: the One-Fifty, the Two-Ten, and the Bel Air.

The One-Fifty Series

The base series of the Chevrolet line was the One-Fifty. There was the two-door sedan, four-door sedan, two-door utility sedan, two-door station wagon, and Handyman wagon (also called the Sedan Delivery). All were six-passenger vehicles except the two-door utility sedan, which had a bench front seat for three passengers with the rear seat removed to create a raised, flat loading area—great for traveling salesmen. The Sedan Delivery had two individual front seats.

Chevrolet published a spiral-bound book, *1955 Chevrolet Features*, for its dealers, illustrating all the features of each series. In this book, the series were referred to as Series 1500, Series 2100, and Series 2400 (Bel Air). The differences in the level of interior and exterior trim, complete engineering features, options, and available paint and interior color combinations were detailed in this book.

Of the 1500 Series, the book had this to say: "The 1500 Series offers four models which, although more conservatively trimmed and less luxuriously appointed than the 2100 and 2400 Series, fully reveal the basic beauty of the 1955 styling with its many new features.

"All of the most important bright metal components of the other two series are also used in the economy models. These include the radiator grille, front and rear bumpers, bumper guards, headlight bezels, hood ornament and emblem, taillight bezels, and the door and deck lid handles. Identifying features include the same front fender nameplate and hub caps as used on the 2100 Series."

The mention of four models in the book was

The 1955 Bel Air two-door sedan sold well, with 168,313 units built. Equipped with a V-8, it sold for less than $2,000. This car is finished in Onyx Black and India Ivory.

This photo is pure 1950s: the car, the fashions, the models, the house. The roofline of the 1955 Nomad was practically identical to that of the Corvette Nomad show car.

an error, as the book illustrated the five models available. Although the interiors were rather spartan, the book described them in grander terms:

"SEDANS . . . The interior treatment of the Series 1500 sedans features cushions and backrests of a new gray pattern cloth which is highlighted by vertical rows of raised chevrons. Contrast is provided by the black elascofab bolsters, and the all-vinyl sidewalls match the seat trim in color and design. Floor mats are textured black rubber.

"STATION WAGON . . . The functional qualities of the 1500 station wagon are enhanced by the interior trim. Available in two color combinations, two-tone green or beige and brown, the vinyl and elascofab interiors are as practical as they are decorative. Featured on the sidewalls, cushions, and backrests is a new textured vinyl, carrying an attractive linked cord pattern.

"Passenger compartment floors are covered by black rubber mats, while the load space uses dark beige ribbed linoleum."

The two workhorses in this series were the

Utility Sedan and the Sedan Delivery wagon. The brochure had this to say about the Utility Sedan: "That's right! This handsome interior is yours in Chevrolet's lowest-priced series. Practical, too, with generous use of long-wearing vinyl trim. And just look at all the extra load space you get in this "One-Fifty" Utility Sedan. The entire rear compartment can be used for hauling cargo—and the floor is raised to make loading and unloading easier."

The Sedan Delivery was not listed in the showroom brochure but was listed in the *1955 Chevrolet Features* book. It was similar in appearance to the two-door wagon, except for the use of stamped sheet metal panels in place of the long rear quarter windows. It was a favorite with painters, plumbers, and other small businesses requiring an inexpensive, unadorned utility type wagon, offering protection from the elements a truck could not offer.

The two-door sedan, four-door sedan, and two-door station wagon made up the bulk of 150 series sales. They were bargains, especially compared to the 1954 models.

For example, the 1954 two-door sedan had a list price of $1,623. The completely new 1955 two-door sedan had a list price of $1,685. The optional V-8 pushed the price to $1,784—only $99 dollars more. The V-8 could be had in the four-door sedan and two-door wagon for the same price.

The One-Fifty Series was available in a plethora of solid and two-tone colors, but there were fewer two-tone colors available, compared to the other two series. Not many buyers opted for the two-tone paint on the One-Fifty Series, and few examples exist today.

The V-8 was a worthwhile option in all the Chevrolet models because of the added oomph it delivered. The trusty six-cylinder was rated at 123hp at 3800rpm. The V-8 put out a whopping 162hp at 4400rpm. However, nearly $100 for the V-8 was no small change in 1955; choosing the optional V-8 for some buyers wasn't always an easy decision. And some buyers simply chose to stay with the six-cylinder engine because of its longevity and known reliability.

The One-Fifty Series Chevrolets certainly had their place in the Division's scheme of things, but production for all five models in the series totaled fewer than 140,000 units for the 1955 model year. Far more popular, in fact the most popular series in the Chevrolet line, was the Two-Ten Series.

The Two-Ten Series

"You can tell just by looking that there's plenty of sports car spirit in this exciting "Two-Ten" 2-Door Sedan. Notice the dramatic curve of the Sweep-Sight Windshield . . . its low "let's go" lines. Imagine—all this glamour in a car that's priced so low!"

So went the foldout showroom brochure, extolling the desirability of the mid-level Chevrolet. Copy like this went a long way to making the Two-Ten Series a tremendous seller.

"Seated behind that wheel, you can see all four fenders. And wherever you sit, you'll enjoy the fashionable two-tone upholstery and smart, new appointments of this beautifully color-keyed interior."

There were six distinct models in the Two-Ten Series: the two-door sedan, four-door sedan, two-door Delray club coupe, two-door sport coupe (introduced in June 1955), two-door Handyman wagon (with three-quarter rear glass instead of sheet metal panels), and four-door Townsman station wagon.

The three pieces of exterior decorative trim distinguishing the Two-Ten from the One-Fifty were the windshield and rear window reveal moldings, rear fender beltline molding, and bright, metal side window sill moldings on the sedans and club coupe.

This trim differed on the station wagons. The diagonal sash molding that intersected the rear fender spear molding was shortened; it did not run all the way up the rear window sill molding. Also, the belt and header moldings ran from the windshield straight to the back and across the tailgate, above and below the window glass.

Of the Two-Ten two-door station wagon, the showroom brochure waxed eloquent: "Chevrolet presents a whole new idea of station wagon usefulness and beauty for 1955! Who could ask for a handsomer car than this low (6 inches lower!) 'Two-Ten' 2-Door Station Wagon! Notice how the side windows curve around at the rear, letting you 'see through' the corners. Look at those long and graceful lines. Then take a peek inside. Here's an all-vinyl interior that's as practical as it is beautiful. And it's beautifully color-keyed to go with whatever exterior you select. Seats six with room to spare—and there's more room with the rear seat folded."

There were more interior color and fabric/vinyl combinations available in the Two-Ten Series. The One-Fifty Series had only one interior color combination for the sedans and two combinations for the station wagons. The Two-Ten Series had three interior color combinations for the sedans and coupes and three for the station wagons.

Thanks to the addition of the rear fender beltline trim molding shared with the Bel Air, the Two-Ten Series was a natural for the two-tone paint treatment, and more two-tone colors were offered on this series. With only a modest price increase over that of the One-Fifty Series, the extra trim and color combinations made the Two-Ten a popular line, selling even better than the Bel Air. Nevertheless, it was the Bel Air that got the lion's share of the press coverage, and they are the most-often photographed cars today.

The Bel Air Series

The Bel Air name first appeared in 1950 as a two-door coupe as part of the Styleline series within the Deluxe 2100 HK line of Chevrolets. It became a distinct line of its own in 1953, offering the highest level of trim in the Chevrolet line in four different bodystyles: two-door sedan, four-door sedan, two-door sport coupe, and two-door convertible. A four-door station wagon was added in 1954. Along with these five models, Chevrolet introduced a new model within the Bel Air line in February 1955—the two-door Nomad. The creation of the Nomad has an interesting history, and its low production volume has made it a collector's classic.

In 1953, Harley Earl informed the Chevrolet studio he wanted a special Corvette created for the 1954 Motorama, using themes from the fast-back Corvette show car, called the Corvair, which would also be shown at the 1954 Motorama. Earl

The Nomad Story

The dictionary defines nomad as "1. a member of a people or tribe that has no fixed abode, but moves about from place to place according to the state of the pasturage or food supply. 2. any wanderer." It seemed an appropriate name for a proposed show car conceived by Harley J. Earl, chief of GM's Styling Section. Earl joined GM in 1927 at the age of thirty-two and enjoyed a rich career there. He quickly "wandered" through GM's hierarchy, gaining pro-minence as he went; by the 1940s, he held sway over the appearance of the General's cars. World War II interrupted automotive production, all but halting work on automotive styling. Even after the war, automotive design seemed frozen in time, and this stylistic lethargy wasn't shrugged off until the early 1950s.

The golden age of auto styling began in 1949 with the first Transportation Unlimited show hosted by General Motors. The show served both as a means of displaying GM's forward thinking and as an opportunity to get a reading on the buying public. As the fifties dawned, Earl looked forward to these shows as an opportunity to produce exciting, ground-breaking showpieces. These show cars often incorporated features that, it was implied, might appear on future production cars. They also displayed styling and features that could not possibly be offered from a production or cost standpoint, but were meant to make the heart beat a little faster.

One of the most exciting programs Earl was working on in the early fifties involved a new sports car, the Corvette, to be introduced for 1953. The Korean War temporarily halted the Transportation Unlimited show, but it returned in 1953 as Motorama. It was at the 1953 Motorama that Chevrolet displayed the new Corvette, and the response was, predictably, overwhelming. The Corvette was all the more stunning because it was essentially a show car that went into production, albeit limited production.

Earl decided to work on two Corvette-based show cars for future Motoramas. One was a fastback version that emerged as the Corvair. The other was a two-door station wagon which retained much of the Corvette's bodywork while having a longer wheelbase and new roofline. This second showpiece, the Nomad would later provide the basis of the production version that eventually emerged in 1955.

When the Corvette Nomad was displayed at the 1954 Motorama, held at the Waldorf Astoria Hotel in New York City, it created a sensation. Chevrolet's interior and exterior studio designers and stylists were working Saturday when the crowds poured into the Waldorf to see the breathtaking show cars on display and view the new production 1954 GM models. Clare MacKichan, Chevrolet interior and exterior design studio chief, was at the studio that Saturday when he received a call from Harley Earl's assistant, Howard O'Leary. The decision had just been made to make a Nomad production car, but based not on the Corvette but on a production station wagon platform from the finalized 1955 car line. O'Leary told MacKichan to get his team together, get to work, and have something to show for the following Monday—only two days away!

The design team pulled drawings for the Nomad along with the drawings for the Handyman two-door station wagon. Assistant Chief Designer Carl Renner carefully grafted the Nomad's roofline with wrap-around rear side glass and sloping tailgate onto the Handyman body. The show car's full rear wheel openings were also retained. Renner was thoroughly familiar with the Nomad's design, as it was based on his concept. Orthographic views (front, rear, sides, and top) of the proposed production version were made by cutting and taping the two sets of drawings together. Carl Renner produced renderings to show the car in color. Practically all the design elements of the show car's roofline and tailgate were retained for the production car, but the body itself could not be changed from the 1955 production model, due to both the prohibitive cost and the desire to retain a family resemblance to the Chevrolet car line.

Among the Nomad's most distinctive stylistic features were the nine ribs that spanned the roof. Earl had originally proposed that the Nomad have a retractable stainless steel roof, operating much like a collapsible cup. The concept was abandoned on the show car, but the ribs remained as a styling element. The chrome trim bars on the Nomad tailgate were adopted for the production car virtually unchanged, as were the vertical tailgate handle and even the Nomad script above the handle. The production car also received chrome "eyebrows" which blended into moldings on the front fenders, sweeping back to the trailing edge of the doors. The Nomad would be, appropriately, part of the Bel Air line.

The 1955 Nomad was introduced in January of that year. Although sales failed to match expectations, they were respectable for a specialty car, at 8,530 units. Chevrolet was far more concerned about Corvette sales, which didn't tally even 4,000 units in 1954. Fortunately, Chevrolet chose to keep both the Nomad and the Corvette in production. While not the sales success Chevrolet had hoped, the Nomad is immensely popular in Tri-Chevy circles today.

What fate ultimately befell the Nomad show car? The car was kept by Chevrolet until sometime in the sixties when it was finally destroyed—the fate of most dream cars once they had served their purpose on the show circuit. Over the years, rumors have surfaced, claiming it was really kept in storage in some GM facility warehouse, but the truth is less intriguing. It does, however, make for interesting speculation.

At the 1954 Motorama held at the Waldorf Hotel in New York, Chevrolet displayed the Corvette Nomad show car. Styling cues, primarily the roofline, would appear on the 1955 Nomad. Note the tailpipe exiting through the rear fender.

wanted, essentially, a Corvette station wagon, but with a more rakish tailgate area. A number of the designers at the studio made proposal sketches, and Earl selected the one done by Carl Renner. The appropriate dimension drawings were made for the show car, called the Nomad, to be built. At the 1954 Motorama, the Corvette convertible, the fastback Corvair, and the two-door Nomad station wagon show car were all displayed. Response to the Nomad was enthusiastic. Earl then conceived of a production version, to be built on the 1955 Chevrolet two-door station wagon platform,

with the key styling elements, the roofline and the tailgate, employed in a new production model. Production was approved, and the Nomad project was given to Renner to deliver the renderings and drawings necessary. Due to previous production and styling commitments, the Nomad was not introduced with the other 1955 models, but made its appearance some months after the official introduction of the other Bel Air models.

Since the Nomad was not part of the 1955 showroom brochure, Chevrolet had its ad agency prepare an ad to tout the car's availability and de-

sirability. "Beautiful Addition to Any Resort! The brilliant new NOMAD. It's the brightest new star on wheels—and a brand-new addition to Chevrolet's fabulous fleet of station wagons! This 'do everything' dreamboat was patterned after Chevrolet's famed experimental Nomad with rakish forward-slanting window pillars and a glamorous 'show car' interior. And with all its beauty, the new Nomad is as useful and practical as you expect a station wagon to be. Drop around at your Chevrolet dealer's and see it the first chance you get."

The Nomad was the most expensive model in the Bel Air line, at $2,571 when equipped with the V-8. The two-door sedan was $1,987, the four-door sedan was $2,031, the sport coupe was $2,166, the convertible was $2,305, and the Beauville four-door station wagon was $2,361 (all prices for V-8-equipped cars).

With the Bel Air being the glamour girl in the Chevrolet line, it got the bulk of the coverage in the showroom brochure. This is what the brochure had to say of the four-door sedan: "Here's Chevrolet's new "show car" styling at its beautiful best. The new Bel Air 4-Door Sedan looks as young as you feel behind the wheel! Wider and longer looking, and much, much lower—its crisp lines enhanced by tasteful chrome accents that distinguish the luxurious Bel Air models."

The Bel Air sport coupe was also described in glowing and vivid terms, designed to pull that checkbook right out of the prospective buyer's pocket: "It's got glamour—and plenty to go with it! The new Bel Air Sport Coupe has a dashing, distinctive beauty all its own. And that goes for the interior, too. Smart straw-patterned cloth is combined with panels of rich-looking vinyl in the luxurious two-tone interiors. And every inch is color-keyed to harmonize with breath-taking Bel Air exterior colors."

The Bel Air exterior appearance was distinguished from the Two-Ten Series by the addition of the Bel Air script and Chevrolet crest on the rear fenders, a broader and more decorative "spear" along the rear flanks, the addition of bright trim on the windsplits and brightwork around the

A long list of dealer-installed options were available. This Nomad is fitted with a prismatic traffic light viewer (left), a helpful accessory when the exterior visor was ordered, and a compass (right).

side windows, and full, polished stainless steel wheel covers.

Inside the Bel Air, the luxury treatment continued. The dash received a bright metal center insert sporting tiny Chevrolet "bow ties" repeatedly die-stamped to give it a textured pattern. A gold-plated "Bel Air" logo appeared on the radio grille, and the dash featured bright metal control knobs. At the base of the radio grille on the passenger side was a standard electric clock.

The steering was a three-spoke design with a gold-plated emblem and a distinctive, bright metal horn ring.

The *Features* book described the Bel Air interior this way: "Particularly distinctive is the sidewall styling. A bright metal molding, in the shape of a flat V on the front door and sweeping rearward, divides the light and dark tone fabrics of the sidewall trim. Center panel material is continued onto the sides of the built-in armrests which are in-

stalled on the door of all models in the 2400 Series except the station wagon rear doors. Eighteen interior trim combinations, divided among the five models in the series, are keyed to the exterior colors. To harmonize with the exterior colors, the interiors are available in two-tones of green or blue, brown and beige, turquoise and ivory, or coral and gray."

In addition to the interior treatment of the sport coupe described earlier, this model received further bright metal in the form of exposed roof bows, rear window molding, interior light bezels, and additional brightwork along the upper part of the side windows. Like in the sedans and the convertible, the floor was fully carpeted.

The convertible interior was designed with the elements in mind. The seats were finished in ribbed vinyl, which Chevrolet called elascofab, and offered in six two-tone color combinations to coordinate with the exterior colors, then set off with white saddle stitching. Additional brightwork was added around the inside of the windshield on the convertible.

The Beauville station wagon interior was described in the *Features* book this way: "Combining outstanding versatility with the distinguishing appearance of other models in the 2400 Series, the station wagon features interiors similar to those of the Sports Coupe. Cushions and backrests are in the new straw pattern cloth with bolsters in either blue or beige leather grain elascofab. All-vinyl sidewalls and headlining complete the interior trim. Colored rubber mats cover the front and rear passenger compartment floors."

How different was the Nomad interior from that of the Beauville? Where the Beauville had only two interior color combinations in cloth/vinyl, the Nomad had seven two-tone all-vinyl color combinations. The unique waffle-like pattern used on the seats was repeated on the doors. Unlike that of the Beauville, however, the Nomad floor was fully carpeted. One of the most distinguishing features of the Nomad interior was the interior headliner brightwork bows that mirrored the exterior ribs on the roof.

The Bel Air line was all the prospective Chevrolet buyer could ask for. There was a bodystyle for practically every need or want and an array of solid and two-tone exterior colors to chose from with striking color-coordinated interiors. However, Chevrolet was able to keep the price low by making so many desirable, and some would say necessary, features optional.

Extra-Cost Equipment

All 1955 Chevrolets came standard with the 123hp inline six-cylinder engine with three-speed manual column-shift transmission. When coupled with the two-speed Powerglide automatic transmission, it was rated at 136hp. The Powerglide option was $178.35. A new overdrive transmission, coupled with the standard three-speed manual transmission with a 4.11 rear axle ratio, was available for $107.60; this combination improved acceleration times considerably while lowering engine rpms significantly when the overdrive transmission was engaged. Then, of course, was the

162hp V-8, a veritable bargain at $99 regardless of model. The V-8 was also available with the overdrive transmission or the Powerglide transmission.

The hot news among enthusiasts, however, was the four-barrel carburetor option available on the V-8. The High Performance Package included the four-barrel carburetor, special intake manifold flanged for the carburetor, dual exhaust system, and high-capacity oil bath air cleaner. All this for the amazing price of $59.20! The V-8's performance was boosted from 162hp to 180hp at 4600rpm with 260lb-ft of torque at 2800rpm.

There were other options that were niceties to have on your Chevrolet. Those living in cool or cold climes would definitely want the Airflow heater, for $72.80. Those living in warmer sections of the country, or those not wanting to endure summers, could fork over $565.00 for air conditioning. A/C was new for 1955, and with a price tag nearly one-third the retail price of the cars themselves, there were not many takers. It was available only with the V-8, and if you ordered this expensive option, you could not order the heater, and vice versa.

Other worthwhile, factory-installed options, to name a few, included "power-positioned front seat" and power windows for $145.30, but both of these could be optioned separately. Power steering was $91.50, power brakes were $37.70, and tinted glass was $32.30.

There was also a staggering array of dealer-installed options. Among them were the Autotronic Eye ($44.25) which automatically lowered the high beams when oncoming cars approached, the Continental spare wheel carrier ($118), pushbutton radio and antenna ($84.50) (the manual radio was $62 and a special Signal Seeker radio was $105.00), dash-mounted traffic light viewer for $2.90, outside windshield visor for $19.90. These only scratched the surface of all the available options your friendly Chevrolet dealer offered.

The introduction of the 1955 Chevrolets was not only the biggest thing to happen to Chevy dealers in years, it was the biggest thing to happen to Chevrolet in years! The Division rolled out the advertising carpet and the Chevrolets were one of the most heavily promoted makes that year, with

Next pages
An air conditioning vent on each side of the dash directed cool air to the driver and passengers. A decal affixed to the vent panes states that the owners of this car travel in air-conditioned comfort.

ads appearing in countless magazines. It was time to beat the drum, and one of the best ways to do that was through racing.

Racing Improves the Breed

Aside from the small-block V-8 itself, the High Performance Package did more to thrust the 1955 Chevrolet into racing—and winner's—circles than just about anything else. This package was promoted by the magazines, indirectly, by road tests. *Road & Track* tested a two-door Two-Ten equipped with one of the 180hp engines. It reached 60mph in a respectable 9.7sec and covered the quarter-mile in 17.4sec. *Motor Trend* was moved to state that the 1955 Chevrolet was ". . . a completely new make of car."

Racers—both amateur and professional—came out of the woodwork to compete with the new V-8-powered Chevrolet. The car did extremely well in its class in various trials. By June 1955, a Chevrolet ad appeared in magazines boasting, "Don't argue with this baby!" The ad went on to recount the class wins by the 180hp "Super Turbo-Fire V-8" at Daytona Beach (where they still raced on the beach) and other NASCAR events.

However, the zenith of hyperbole was the ad that appeared in the September 1955 *Motor Trend*. No one could accuse Chevrolet of not beating its own drum: "Chevrolet's got it! Enough high-powered punch to run the pants off the competition—all competition, including most of the so-called 'hot' high-priced cars!

"If you've seen a '55 Chevrolet in action this news doesn't surprise you. You've witnessed the swivel-hipped way it handles, breaking through the pack to come lane-hugging through the turns—wide-spaced rear springs holding tight—then digging out with a catapult surge of V-8 power! If you haven't—brother, it's something to see. Better yet, put a new Chevrolet through your own paces. Your dealer has one waiting . . . "

All hyperbole aside, Chevrolet did have reason to crow. Despite a crowded V-8 market, the "leader in the low-priced field" managed to make a big name for itself, and it used racing to do so. Chevrolet kept the racing-win-inspired ads coming: "CHEVROLET'S TAKING COMPETITION

TO THE CLEANERS!" This ad touted Chevrolet's record in NASCAR short-track events, stating Chevrolet had won twice as many races as the nearest competitor. "CHEVROLET WINS AT ATLANTA!" referred to the 50-mile NASCAR event meant to extract a car's ability to accelerate, handle, and brake under grueling conditions, and Chevrolets placed first and second.

The apex of racing success was the Southern 500 at Darlington, South Carolina. On Labor Day 1955, Herb Thomas in his Motoramic #92, took first place, surviving a field of sixty-two cars. This really put the fire in The Hot One from Chevrolet, and the automotive press went wild.

All this racing activity and success was not entirely by design on Chevrolet's part, but it certainly was no accident, either. Ed Cole saw the outstanding engineering features of the '55 validated by this racing activity, and if racing could help sell Chevrolets, so much the better.

The new 1955 Chevrolet was an engineering and styling triumph. The Division successfully changed the image of the car from being an older man's car to one with a more youthful image. Market reluctance to buying an all-new car was also overcome. Chevrolet sold 1,414,365 passenger cars in 1954. When the 1955 sales numbers were tallied by *Ward's*, an automotive industry journal, Chevrolet had sold 1,830,038, compared to Ford's 1,764,524. Chevrolet was still the sales leader, and it looked like The Hot One was going to get even hotter in 1956.

CHAPTER TWO

Turning Up the Heat

Nineteen Fifty-Six was a year of headlines: Nikita Khrushchev denounced Stalin's excesses, the first aerial test of a hydrogen bomb equivalent to 10 million tons of TNT took place over Bikini Atoll, a workers' uprising against Communist rule in Poznan, Poland, was crushed, Egypt took control of the Suez Canal, war broke out in the Middle East as Israel invaded Egypt's Sinai Peninsula, British and French troops invaded Egypt at Port Said, a Middle East cease fire was called, and an anti-Communist revolt in Hungary was crushed by Soviet troops and tanks.

MacKinlay Kantor won a Pulitzer Prize for his novel, *Andersonville*. Yul Brynner won an Oscar for Best Actor for his role in *The King and I,* Ingrid Bergman won Best Actress for her portrayal in *Anastasia*, and Best Picture was awarded to *Around the World in 80 Days.*

Americans found escape is such popular TV programs as "Father Knows Best," "The Millionaire," "Make Room for Daddy," and "The Dinah Shore Chevy Show." The hottest book was—believe it or not—*Arthritis and Common Sense,* with more than 250,000 copies sold.

Among the songs most listened to were "Don't Be Cruel" and "Heartbreak Hotel" by Elvis Presley, "Whatever Will Be,

Will Be" by Doris Day, and "Great Pretender" by The Platters.

To launch its 1956 car line, Chevrolet's advertising agency used the tag line, "The Hot One's Even Hotter!" Despite almost completely new styling for 1956, the emphasis in nearly all ad copy was on the car's ability to accelerate for safe passing, its handling, and its superior braking.

Because the 1955 model was completely new and the 1957 version is perhaps the favorite among Tri-Chevy enthusiasts, the 1956 car is sometimes overlooked and seen instead as a mere facelift car. Facelifted it was, but to some it is the

If anything, the Nomad looked even better for 1956. Since the Nomad entered production late in the 1955 model year, this was the first full year of production for Chevrolet's "sportswagon."

Loves to go...and looks it!
The '56 Chevrolet

The Bel Air Sport Sedan is one of two new Chevrolet 4-door hardtops. All 19 new models feature Body by Fisher.

It's got frisky new power... V8 or 6 ... to make the going sweeter and the passing safer. It's agile ... quick ... solid and sure on the road!

This, you remember, is the car that set a new record for the Pikes Peak run. And the car that can take that tough and twisting climb in record time is bound to make *your* driving safer and more fun.

Curve ahead? You level through it with a wonderful nailed-to-the-road feeling of stability. Chevrolet's special suspension and springing see to that.

Slow car ahead? You whisk around it and back in line in seconds. Chevrolet's new high-compression power—ranging from the new "Blue-Flame 140" Six up to 225 h.p. in the new Corvette V8 engine, available at extra cost—handles that.

Quick stop called for? Nudge those oversize brakes and relax. Chevrolet's exclusive Anti-Dive braking brings you to a smooth, *heads-up* halt.

No doubt about it, this bold beauty was made for the road. Like to try it? Just see your Chevrolet dealer.... Chevrolet Division of General Motors, Detroit 2, Michigan.

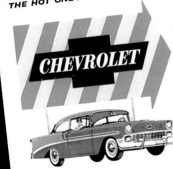
best-looking of the three model years. There were also some significant firsts that made their debut for 1956.

One has to remember that Ed Cole had instituted a tremendous increase in the Chevrolet staff for the 1955 car. Once that was completed, the engineers, designers, and stylists were put to work on the '56. And there was much that was new about the 1956 Chevrolet. Chevrolet Motor Division published *Finger-Tip Facts About the 1956 Chevrolet*. It stated:

"Whether a person is young or old, he will thrill to the exciting longer lines and gay new colors of the 1956 Chevrolet. Whether he is a driver or passenger, he will enjoy the good taste expressed in the smart new styling and colorful new fabrics of the car's interior. Whether he has driven one mile or a million, he will experience new driving pleasure when commanding any of Chevrolet's improved, efficient, and powerful engines. And, whether he is well-to-do or of moderate means, he will be glad to know there's a car intended for him in Chevrolet's greatly expanded line.

"With its fresh, eager, champing-at-the-bit look, every new Chevrolet stands out as an example of tasteful, modern styling. Without exaggerated massiveness and bolted-on decoration, it is simply beautiful. In its lovely and gracious two-tone interior, there is luxurious sitting-room comfort for all passengers, and there's plenty of room for their luggage as well. Superb vision gives an outdoor living-room effect."

New models for 1956 included the four-door hardtop Bel Air Sports Sedan, the four-door hardtop 'Two-Ten' Sport Sedan, Bel Air Beau-ville nine-passenger station wagon, and the 'Two-Ten' Beauville nine-passenger station wagon. There were now 364 model/color combinations available, and thirteen of the fifteen exterior colors were new for 1956. Interiors were all new in design, colors, fabrics, and vinyls. Output of the Blue-Flame 140 six-cylinder engine, Turbo-Fire V-8, and Super Turbo-Fire V-8 were all increased for 1956. Clearly, Ed Cole and his team had been busy.

All-New Styling for 1956

Most apparent to prospective buyers was the 1956 models' new styling. Except for the roofline, practically every piece of stamped sheet metal was new. Incredibly, the cost to the Division to restyle and retool the 1956 line was $40 million! And this was in 1950s dollars! What did *Finger-Tip Facts* have to say about it? "From the front, there's a new sense of road-hugging stability about the 1956 Chevrolet . . . a broader, lower, stronger look that imparts big car dignity as well as a challeng-

ing eagerness to take off and go."

There had been some internal controversy concerning the grille design of the 1955 model. Some argued the egg-crate grille of the 1955s were reminiscent of the Ferraris and lent the cars a touch of sports car flair. Harley Earl himself was the chief proponent of the 1955 grille design. However, the 1955 grille did not meet with universal acceptance within Chevrolet while the car was taking shape. Some felt the grille was too narrow. Another proposed grille design ran full width, was somewhat lower, and incorporated larger rectangular parking lights. The front end design that emerged for the 1956 model year incorporated most features of the alternate grille design origi-

Previous page
The theme for 1956 was "The Hot One's Even Hotter!" This ad emphasized the fact the 1956 Chevrolet "Loves to go . . . and looks it!" The copy in this ad is particularly worth reading; its message really isn't much different from that used in current Chevrolet ads, although today's wording might be a bit more sophisticated. The Bel Air Sports Sedan featured here is shown in Adobe Beige and Sierra Gold.

The interior dimensions of the 1956 Chevrolet were generous.

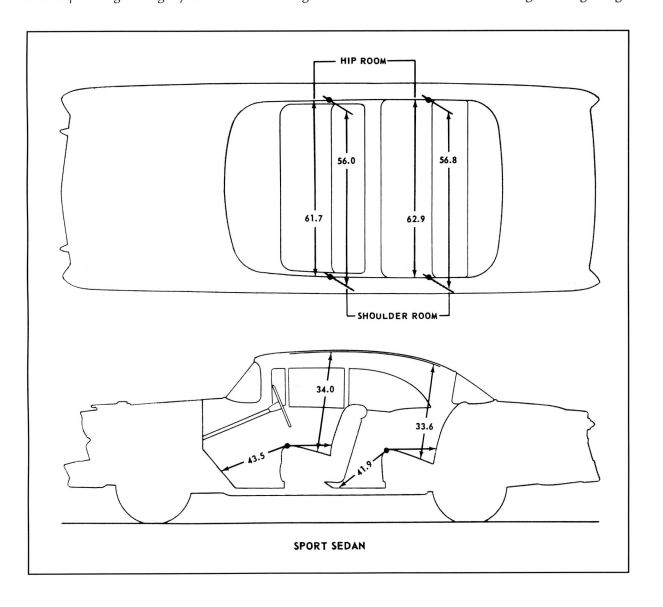

SPORT SEDAN

This jig held various body panels and the roof in place while they were welded together. The body taking shape is that of either a Two-Ten or a Bel Air four-door Sport Sedan.

nally conceived for the 1955 model.

For 1956, the hoods surrounding the headlights were squared off and were more prominent. The hood itself was extended several inches. The front bumper was unchanged, but the vertical bumper guards were redesigned and had a more forward rake. The front fenders had a recess, and the crease was carried to the top of the wheel openings. The trim molding that appeared on the Two-Ten and Bel Air was also new and now swept in a graceful curve toward the rear bumper. The trim molding for the Two-Ten differed from the Bel

Air and the two-tone paint schemes for the two models also differed as a result.

Interestingly, the One-Fifty model also sported side trim similar to that of the Two-Ten, but instead of sweeping down to the rear bumper, the molding was straight and stopped before reaching the rear wheel opening. A diagonal portion of the trim near the end of the molding was carried up to the rear passenger window, and this permitted two-tone paint schemes on the One-Fifty for the first time.

The rear taillights differed for 1956 as well. The redesign incorporated a fuel filler concealed behind the left taillight. Chevrolet noted, "Not only does this design avoid the beauty-marring effect of an external door in a fender or an exposed

Once the main body was welded together, a body finish man would fill any exposed seams with molten lead. Once cool, the leaded seam would be filed and sanded before the body was primed.

One of the final steps in producing a rolling chassis was the mounting of the wheels and tires. This was done using a five-position pneumatic lug wrench.

cap, but it also places the filler high for easier filling." It also appealed to car buyers with a fascination for clever design. The rooflines and glass areas for 1956 models remained unchanged, but the new four-door sport sedans did have unique rooflines, rear window, and side glass.

How did this restyling effort affect the

dimensions of the 1956 Chevrolet? Wheelbase remained unchanged at 115in. Overall length grew from 195.6in to 197.5in. Station wagons grew from 197.1in to 200.8in. The rooflines and glass area of the station wagons and Bel Air Nomad did not change.

The way the 1956 Chev-rolet was described in *Finger-Tip Facts* was contradictory to say the least. It referred to its "massive new front-end styling" and "The emblem for the six-cylinder cars is broad, in keeping with the sense of massiveness in the car . . . " At the same time, Chevrolet extolled the car's "new rearview fleetness," stating, "From the rear, as well as from the front, the 1956 Chevrolet

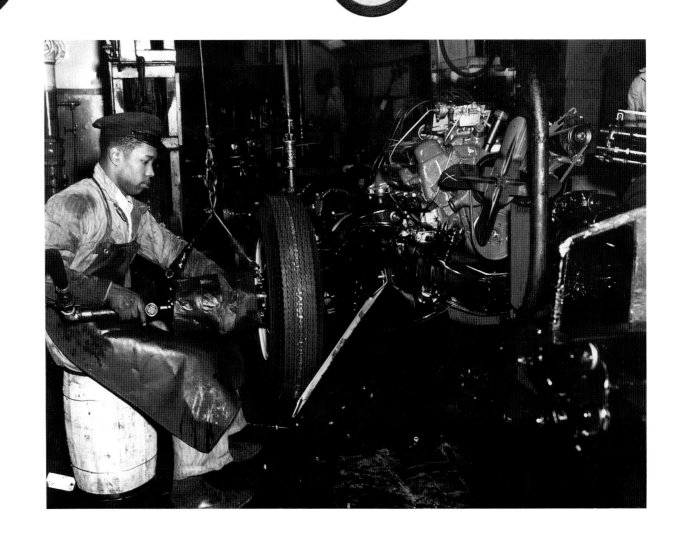

coveys an impression of even greater massiveness and fleetness." Surely some dealers were puzzled to learn from the manual that the 1956 Chevrolet maintained a "compact exterior"; as a result, " . . . maneuvering in today's multi-lane traffic is easier with the Chevrolet than with cars having larger exteriors and the Chevrolet can be parked in spaces that are too short for cars that are less compactly designed." The 1956 Chevrolet was truly an automotive oxymoron.

"Revolutionary Design"

Naturally, Chevrolet had to sell the new 1956 line to its dealers and the dealers to their prospective customers. This was where *Finger-Tip Facts* came in. It educated the dealer in general and the salesmen in particular on what was new. In truth, 1955 was the year of radical change and improvement for the car line. Nevertheless, that was last year. Chevrolet was stating that the changes to the 1956 car line were every bit as dramatic:

"The 1956 Chevrolet is revolutionary. Not only does it outclass all cars in its field, but it steals the thunder from higher-priced cars in features, styling, utility, and ability.

"Nearly every person considers his automobile to be one of his proudest possessions. He may admire it for any or all of its many fine features

The nine-passenger station wagon was an efficient mode of mass transportation. The two rear seats could be folded to carry cargo or a combination of passengers and cargo.

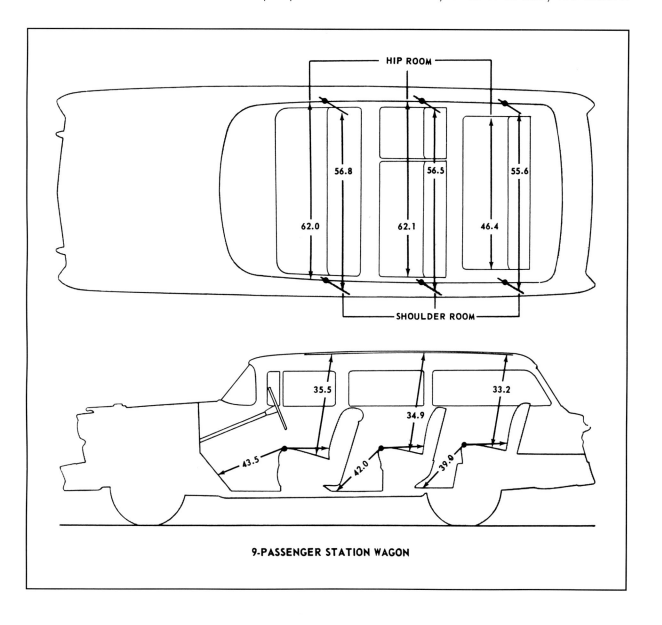

9-PASSENGER STATION WAGON

The restyling effort by the Chevrolet studio for the 1956 model year was most successful. Chevrolet managed to combine, in its own words, "massive new front-end styling" with "new rearview fleetness." This Bel Air Sport Coupe, finished in India Ivory and Nassau Blue, is owned by Dennis and Mary Schrader.

but, when he considers why he bought it, two bedrock, fundamentals stand out. These are the car's utility and ability: utility in the sense of the room it provides for him and his passengers, ability in the sense of its performance. Some cars are big—with a sacrifice in ability; others are small—with a sacrifice in utility. In the 1956 Chevrolet, however, sound engineering has resulted in a car that has ideal proportions—in size, weight, and power—to provide an ideal combination of utility and ability. Not only is this car big inside for comfort, but it's compact outside for enviable maneuverability and it's loaded with power for record-breaking performance.

"This ideal combination stems from the car's

compact design. Compactness in the sides broadens the interior, yet decreases the car width. Compactness in the chassis and body makes the car low for better roadability and appearance, with a low hood and deck that make the road immediately ahead and behind the car easy to see. Compactness ahead of the front wheels and behind the rear wheels keeps the overhang small and avoids excessive length that makes maneuvering hard. For stability, not only is the center of gravity low, but the wheel treads and spring centers are broad in relation to the car width. Likewise, the wheelbase is long for the car length. Compact design also pares dead weight to the bone, and with Chevrolet's great power, provides a very low

THE BEL AIR SPORT SEDAN—One of Chevrolet's two new 4-door hardtops. Body by Fisher, of course.

Surest cure for Spring Fever— a fresh and frisky Chevrolet!

Maybe you'd like a frisky four-door hardtop—or a high-fashioned

station-wagon, perhaps? Chevrolet has them all. Convertible? Sedan? Sports Coupe?

You'll find your heart's desire in the 20 lithe and lovely models for 1956.

A General Motors Value

'56 CHEVROLET

Chevrolet traded the beauty-marring effects of a rear-fender fuel filler door for a gas cap hidden behind the hinged left taillight.

weight-to-power ratio."

Chevrolet was trying to be all things to all prospective buyers, and in many respects it succeeded. Proof of its success was its number one sales position. Unlike today's diverse product line, in 1956 Chevrolet had but one basic design upon which to base its appeal. There were two-door and four-door sedans, two-door and four-door station wagons, two- and four-door coupes, the rakish two-door Nomad, a convertible—quite an array of cars. If you couldn't find what you needed or wanted from that selection, you were probably in the market for something completely different.

New Exterior Colors

For 1956, Chevrolet all but threw away the palette. Thirteen of the fifteen colors were new for '56. The two colors carried over from 1955 were Onyx Black and India Ivory. There were ten solid

car colors and fourteen two-tone color combinations. Eight of the solid colors were new for 1956. The five additional new colors were part of the two-tone offerings.

The ten solid car colors were Onyx Black, Crocus Yellow, Sherwood Green, Pinecrest Green, Harbor Blue, Nassau Blue, Matador Red, Dusk Plum, India Ivory, and Twilight Turquoise.

The fourteen two-tone combinations included India Ivory and Onyx Black, Onyx Black and Crocus Yellow, Crocus Yellow and Laurel Green, Sherwood Green and Pinecrest Green, India Ivory and Sherwood Green, India Ivory and Pinecrest Green, Nassau Blue and Barbor Blue, India Ivory and Nassau Blue, India Ivory and Dusk Plum, India Ivory and Matador Red, Dune Beige and Matador Red, Adobe Beige and Sierra Gold, India Ivory and Dawn Gray, and India Ivory and Twilight Turquoise.

The Schraders' car is a fine example of a few-frills Bel Air. The majority of Bel Airs built in 1956 were ordered with optional full wheel covers, but this one was delivered with the standard small wheel covers.

Racing Improves the Breed

The 265ci small-block V-8 transformed Chevrolet's "stovebolt" image, turning it into General Motors' racing Division. And when Chevrolet entered racing in earnest, it revolutionized motorsports in general. Certainly, within Chevrolet, Zora Arkus-Duntov knew the performance potential of the small-block V-8 and fully intended to exploit it. Outside Chevrolet, countless independent racers saw the makings of a winning race car in the new Chevrolet and the performance options available with the V-8.

The 1955 Chevrolet made its racing debut at Daytona Beach in February of that year. Cars were still raced on the beach in those days—Daytona International Speedway lay some years in the future. This was a NASCAR Grand National event with a 4.1-mile oval course, with half the course on the beach and the other half on the road that paralleled the beach. In a field numbering more than eighty cars, Jim Radtke in his Chevrolet succeeded in placing tenth. There was more to NASCAR racing than the Grand National events, however. Chevrolet really excelled in smaller events, which were equally important in advertising.

At Daytona Beach in 1955 in the NASCAR Acceleration Tests Over Measured Mile from Standing Start, Chevrolets captured the first four positions in their class and eight out of the first eleven positions. In the Straightaway Running Test for cars delivered for $2,500 or less, Chevys took the first two positions and seven out of the first eleven finishes. In the Two-Way Straightaway Running Over Measured Mile for cars with displacements between 250 and 299ci Chevrolets took three of the first five places.

The small-block V-8-powered Chevrolet did very well in NASCAR short-track events in 1955. On the half-mile track at Columbia, South Carolina, in the NASCAR 100-mile event, Chevrolet finished first. At Fayetteville, North Carolina, Chevrolet finished first in the late Model Event. At Atlanta, Georgia, Chevrolet finished first and second in the 50-mile event. In total, Chevrolet won more points in Short Track events in 1955 than any other make.

One of the biggest racing events of 1955 was the Southern 500 held at Darlington, South Carolina. Out of a field at sixty-nine cars, there were twenty-four Chevrolets racing for the gold. Herb Thomas, piloting a Motoramic V-8-powered Chevy massaged by Smokey Yunick, claimed the checkered flag with seven of the first ten finishers also piloting Chevrolets. No one questioned how hot The Hot One truly was.

When Zora Arkus-Duntov established his Pikes Peak records in a disguised 1956 Chevrolet in September 1955, the Division made sure it ran a four-page ad detailing every aspect of the Peak assault. Less disguised was the obvious performance message, though it was couched in safety rhetoric: "No other car has ever gone so high so fast—so safely!"

By the end of the 1955 season, Chevrolet had amassed 668 points in short-track racing events. Oldsmobile was a distant second with 195 points and Hudson third with 184 points. Dodge was in fourth place with 176 points, and Ford trailed in fifth with 165 points. This was an embarrassment to Ford, and the company turned up the heat for the 1956 season. Still, by the end of April 1956, Chevrolet remained clearly dominant in short-track racing having accrued 226 points to second place Ford's ninety-one points. As a matter of corporate pride, Ford could not let Chevrolet walk away with all the marbles again, and the competition between the two automotive giants really heated up in 1956.

In 1956, Chevrolet's small-block was at a clear displacement disadvantage on the Grand National circuit against such heavy hitters as the 340hp Hemi-head Chrysler 300. Chevrolets won only three Grand National races that year. However, in USAC, Chevrolet won eleven of the race events, while Ford won only four.

USAC, not AAA, was now sanctioning the Pikes Peak Hill Climb. The records established by Zora Arkus-Duntov stood for only one year. Jerry Unser broke Arkus-Duntov's record by more than one minute, racing to the top in sixteen minutes and eight seconds in a 1957 Chevy.

The V-8 Chevrolet also set another record of a different sort. A 1956 Two-Ten sedan, one of four prepared by Smokey Yunick, lapped the Darlington track for twenty-four hours, making only twenty-one pit stops, and averaging 101.58mph. This broke the Production Car class record, formerly held by Chrysler, by more than 11mph.

By 1957, Chevrolet's star was beyond rising—it was the brightest in the automotive racing heavens. Chevrolet introduced its Ramjet Fuel Injection on the enlarged small-block V-8, now 283ci. Ford had its own ace up its sleeve, namely supercharging. NASCAR promptly banned both of these power boosters.

The greatest blow to U.S. motorsports, however, was the ban on factory racing issued by the Automobile Manufacturer's Association which took effect June 1957. Ford's new CEO, Robert McNamara, stated that Dearborn would follow the letter of the ban. General Motors in general and

Chevrolet in particular did not. Chevrolet continued to supply high-performance parts and support to teams racing Chevrolets while Ford performance bits and factory assistance dried up.

Prior to the AMA ban, Fords had won fourteen races to Chevrolet's thirteen. More Chevrolets finished in the money and point standings than Fords, however, so Chevrolet took the short-track championship with driver Jim Reed.

In the Convertible Division, the same come-from-behind trend continued. Of the thirty-seven races in the 1957, Fords and Mercurys won twenty-seven, and Chevrolets won only nine. But when all the points were tallied for finishers, Chevrolets carried the day. Clearly the AMA ban undercut Ford and helped Chevrolet. But as one sage put it, if you can't advertise your wins (a provision of the AMA ban), what's the point?

Still, Chevrolet's small-block V-8 gave motorsports racing in general a tremendous boost during the fifties, a trend which positively exploded during the sixties. And it all had its birth during the Tri-Chevy era, when The Hot One ruled the race tracks.

The ad, "You get the winning V-8 in Chevrolet" offered a lot of information regarding Chevrolet's small-block V-8. It listed six distinct advantages that were not available in any other V-8 on the market, according to the ad. This ad appeared in the August 1956 issued of *Holiday* magazine.

1. IT'S REWRITING THE RECORD BOOKS! This is the V8 that broke the Pikes Peak stock car record. It's the V8 that won both the acceleration and flying mile contests for its class at Daytona Beach—and set a new 24-hour competition track record at Darlington, S. C.

2. BIG WINNER IN COMPETITION! In NASCAR Short Track stock car competition, Chevrolet V8's started this season by winning six of the first six events they entered —against cars of all power and price ranges! That's proof of performance that means safer, happier highway driving.

You get the winning V8 in Chevrolet!

3. NEEDS ONLY FOUR QUARTS OF OIL! Most other V8's require five quarts or more! Because of Chevy's super-efficient, low-friction V8 design, you save on every oil change.

THE HOT ONE'S EVEN HOTTER!

CHEVROLET

America's largest selling car— 2 million more owners than any other make!

4. REQUIRES LESS COOLANT! Here's another measure of efficient engine design. Chevy V8's are cooler running, so they don't need as much water or antifreeze.

HERE ARE SIX IMPORTANT REASONS WHY IT'S THE MOST MODERN V8 OF THEM ALL!

You don't get these six advantages in any other V8 built today. And there's something else you miss out on in the others. It's the pure pleasure of bossing around a car with Chevy's rapid-fire reflexes— plus the solid feeling of stability you have at the wheel. Chevrolet is famous for its roadability and sureness of control. Want a sample? Just see your Chevrolet dealer. . . . Chevrolet Division of General Motors, Detroit 2, Michigan.

5. HIGHEST HORSEPOWER PER POUND! A Chevrolet V8 pours out more power per pound of engine weight than any other V8 going! That not only means livelier performance—it's a good indication of its efficiency and precision.

6. SHORTEST STROKE IN ITS FIELD . . . ULTRA HIGH COMPRESSION! The stroke is actually smaller than the bore! The result is less piston travel and less friction. Another reason for Chevy's red-hot V8 performance is ultra high compression—as high as 9.25 to 1! Nobody spared the "horses" either. Horsepower ranges clear up to 225 in this record-breaking new Chevrolet.

3'

The two-tone color combinations were offered in several different stylings based on the models. The conventional two-tone scheme on the One-Fifty and Two-Ten models was a roof and body in different colors. Then there was the Speedline two-tone color styling. On the One-Fifty models, this was made up of one color above the molding, including the hood, and the complementary color on the roof and trunk and below and behind the molding. On the Two-Ten models, Speedline styling appeared as one color on the roof and below the body molding, and the complementary color along the molding and on the hood and trunk. On the Bel Air models, the Speedline color styling included one color on the roof and trunk and within the area defined by the body side moldings, and the complementary color outside the bodyside molding and on the hood.

Dazzling New Interiors

Finger-Tip Facts made much of the new colors, patterns, and textures for each model in the 1956 Chevrolet line: "Comfortable, convenient, beautiful . . . and fashionable aptly describe the interior of each Chevrolet for 1956. For every Chevrolet provides its occupants with all the comforts and conveniences they may want—as standard, optional, and accessory equipment—in a beautiful new interior styled in the latest fashion. In these gorgeous new interiors, striking new designs are

offered in rich new fabrics—in new two-color treatments for each series . . . And to make the interiors even more cheerful, the lighter color of each combination predominates."

The interior and exterior color combinations among the three series for the nineteen models in the 1956 Chevrolet line were mind-boggling. There were 162 model-color combinations using the ten solid exterior colors available. When the two-tone exterior colors were added, there was an astounding total of 364 model-color combinations!

The instrument panel form of the three models was changed little for 1956, but the decorative treatment—the brightwork—was modified. This was most apparent on the Bel Air; the lattice work panel that ran across the instrument panel no longer carried the repeated bow tie theme, replaced by a pattern similar to that of the car's grille.

What strikes enthusiasts new to the classic Tri-Chevys is the absence of plastic on the instrument panel—a common trait of cars of the era. Virtually everything before the driver was either stamped or cast metal and finished accordingly. The rule of thumb was, "if it's sheet metal, paint it. If it's cast, plate it." Brightwork was an easy means of upgrading the appearance of the interior from one model to the next.

Instrument panel conveniences abounded in 1956. These went beyond the standard or accessory features like the clock, cigarette lighter, or radio. How about a shaver? "A boon to the man who must travel early and late is the accessory four-head electric shaver that plugs into the cigarette lighter socket. It operates on twelve volts D.C. in the car and on 110 volts A.C. at home," stated *Finger-tip Facts*.

One couldn't overlook the performance aspect of the 1956 car line amidst all the "longer, lower, and wider" hoopla. Promoting Chevrolet's performance and racing successes was very much a part of product promotion that year.

Even Hotter Performance

Chevrolet's advertising theme for 1956 was "The Hot One's Even Hotter." Chevrolet public relations, as well as marketing, wanted to promote that theme in a visible way. During the summer of 1955, a number of ideas were discussed both at Chevrolet and the Division's advertising agency, Campbell-Ewald. Austin Chenley of Campbell-Ewald suggested an assault on Pikes Peak with a pre-production 1956 model. Chenley contacted Walt MacKenzie, director of public relations engineering, and offered the idea. MacKenzie immediately liked it, and he called Zora Arkus-Duntov to come to his office to discuss a Chevrolet promotional idea. Arkus-Duntov thought it was a splendid idea and wanted very much to be involved.

Among the issues discussed in August 1955 was the matter of a sanctioning body. At the time, AAA (the American Automobile Association) was the sanctioning body for Pikes Peak record attempts, but a chief officer of AAA, Col. Harrington, had close ties with Ford. This would not do. There were also rumors flying that AAA was looking to get out of the sanctioning business and that the United States Auto Club, or USAC, would be the new sanctioning body. What to do?

The National Association of Stock Car Auto Racing (NASCAR) appeared to be the only logical remaining choice. Bill France, NASCAR's president, was contacted and confidential meetings were held, at which NASCAR was asked to sanction the record attempt. The Broadmoor Hotel in Colorado Springs, Colorado, served as record-setting headquarters. A 1956 Chevrolet equipped with a 205hp Super Turbo-Fire 265ci V-8 fitted with a single four-barrel carburetor, hotter

The production front end of the 1956 Chevrolet was disguised with a bogus front clip because the record attempt was made in September 1955 before the new model was introduced.

On Pikes Peak, a NASCAR official surveys the tortuous 12.42-mile course. The car was further disguised by using a black and white paint scheme.

On September 9, 1955, Zora Arkus-Duntov set a new Pikes Peak record of 17min, 24.05sec using a production 205hp Super Turbo-Fire V-8-powered Chevrolet. GM wasted no time in promoting the victory, encouraging prospective buyers to "Drive the Pikes Peak Record Breaker Today!"

Workers lower a 265ci small-block V-8 into the chassis. Once bolted into place, the driveshaft with universal joints was bolted between the engine and differential. Note the road draft tube that runs from just below the distributor, behind the right cylinder head, and down the side of the engine. This would eventually be replaced by positive crankcase ventilation.

camshaft, and dual exhaust was the weapon of choice. A fiberglass front end and bogus grille disguised the as-yet-unveiled production front end.

Arkus-Duntov organized the entire effort. To

determine the proper suspension setup and gearing necessary, he made the practice runs himself with his wife Elfi handling timing duties. The parts the car needed were shipped to Manitou, Colorado, where GM had a facility equipped with hoists. The necessary suspension parts and higher numerical rear axle ratio, which would become a Regular Production Option, were installed. On September 9, 1955, Duntov brought two-door and four-door pre-production 1956 Chevrolets with identical 205hp engines to make the assault on the 14,000-plus-foot Pikes Peak. The first run took place just after 7:00 AM to take advantage of the cool morning air. Less than an hour later, the second car blitzed the Peak. Both cars broke the twenty-year-standing record of 19min, 25.7sec set by Bus Hammond in 1934, with the two-door Chevy setting a new record at 17min, 24.05sec. In-

All three Chevrolet models could receive this new decorative fender ornament for 1956.

deed, The Hot One was even hotter for 1956!

The Chevrolet literature documenting the Pikes Peak success emphasized the safety aspect of the 205hp 1956 Chevrolet's performance. "No Other Car Has Ever Gone So High, So Fast—So Safely!" It went on to say, ". . . it took this record to prove conclusively that outstanding performance means greater safety . . . and it's built right into this new Chevrolet. It showed beyond doubt that the 1956 Chevrolet guarantees better passenger safety with faster acceleration for quicker passing . . . easy, smooth cornering ability . . . greater stability for smooth, low, road-hugging ride."

None of this product promotion, however, could have been possible without the high-revving 265ci V-8. Chevrolet just didn't pursue this line of "performance" advertising prior to 1955. The engine really did put Chevrolet on the performance map, and Cole wasted no time exploiting the small V-8's capabilities. Racing successes helped the Division increase market share—which was the whole point of all this activity.

One venue to turn up the heat for 1956, aside from advertising, was Chevrolet's own product literature, specifically that directed toward its deal-

Despite the fact that the V-8 was only $100 more, the original buyer of this Bel Air felt the "Blue Flame 140" inline six-cylinder engine was perfectly adequate. It was a tried-and true design, having been in production more that forty years. The compression ratio was up to 8.0:1, which bumped horsepower to 140.

The 1956 Bel Air Sport Coupe was offered with only two interior color combinations: Ivory Vinyl with Charcoal Gray Pattern cloth (shown here in the Schraders' car) and Copper Vinyl with Tan Pattern cloth.

ers. Racing was again the theme used to promote the production cars. Chevrolet prepared literature for dealers stating, "Stock Car Racing Gives You More Opportunities to Flag Down Sales." The literature told dealers how to capitalize on Chevro-

let's NASCAR short-track success.

In addition, Chevrolet offered a special kit of promotional material to help dealers tie in with local stock car racers; this included trunk and door banners with the dealer's name, two different newspaper ads, publicity releases, and radio spots.

When prospective buyers of the 1956 models wandered into the Chev-rolet showrooms to see what all the hoopla was about, they learned that V-8 availability was expanded for 1956. To begin with, there was the two-barrel version Turbo-Fire V-8 that developed 162hp at 4400rpm in the manual

shift car and 170hp at 4400rpm with the Powerglide automatic transmission. Also, the 265ci V-8 for 1956 now featured filtered oil, incorporating a replaceable filter canister. Fred Sherman, who worked in the dynamometer room on the smallblock V-8's development, described the performance improvements made for 1956.

"In 1956," Sherman said, "the cam lift was increased from a 0.332- to a 0.365-inch lift. There was an accompanying increase in horsepower. The four-barrel setup then went to a 9.25:1 compression ratio. It was advertised at 205hp. There was another option released. The standard single four-barrel Power Pak was called RPO 410. There was an RPO 411, which had two four-barrel carburetors mounted on the intake manifold of the 265 V-8. This gave a power boost to 225hp and it was a very fast-accelerating automobile. The linkage on the carburetors was progressive. The rear carburetor had the choke on it. This option was carried over to the 283 for 1957."

Louis Cuttitta, Chevrolet's resident carburetor expert, was called upon to do a number of special projects relating to fuel delivery on performance engines. One project involved not Rochester carburetors on the small-block V-8, but Carter carburetors.

"At the time," says Cuttitta, "four-barrel carburetors were coming along, but Carter four-barrels were superior to anything else. Henry Bowler of Carter Carburetors designed the pads on the intake manifold for the Carter carburetors, and there was no way you could ever bolt on two Rochester carburetors. I worked on the Muncie 265ci boat engine that ran on alcohol. They needed somebody to develop a pair of alcohol-running Carter carburetors on the 265 V-8 that Bill Muncie would run on the Detroit River. I got the job of developing those carburetors, along with a guy by the name of Phill Burr, who ran the dynamometer lab in the GM building. The dynamometer lab at General Motors—this was long before the Tech Center—was on the seventh floor of the building, I believe, and the whole building vibrated when the dyno engines ran. They wouldn't let us run the engines during the day, so we had to run them at night. We stuck the exhaust

The design theme of the 1955 dashboard was carried over for 1956, but the decorative trim in the center was new. The electric clock became standard in the Bel Air. The tissue dispenser remained an option.

pipes out the window in the direction of Henry Ford Hospital. It was exciting because the police would come and the fire department, because occasionally we'd shoot balls of fire out the exhaust pipe at the hospital, and people did not really appreciate that!"

Zora Arkus-Duntov followed up his Pikes Peak success by traveling to Daytona Beach in February 1956. He wanted to extract the maximum horsepower he could from the new V-8, install it in the lightest production Corvette he could build, and attempt to set a new record at the Beach for the flying mile and prove the Corvette could be a 150mph automobile.

Duntov selected the 411 option package to be installed in the Daytona Beach Corvette with a special camshaft he designed himself to boost the engine's power even further. The Duntov cam pushed horsepower to 240. To cut drag, he had a special driver-side-only windshield installed on

Chevrolet pursued high-performance engine development of the small-block V-8, and its racing successes were often an advertising tool. No attempt was made, however, to announce what was under the hood. This Bel Air Sport Coupe was ordered with the dual four-barrel carburetor induction system V-8, but the car displayed no outward sign that it was the ultimate performance 1956 Chevrolet.

the car, and a tonneau cover was fabricated to cover the passenger side of the compartment to smooth air flow. At GM's Mesa Arizona Proving Grounds, Duntov pushed the car to over 160mph! On the beach that February, he achieved an average for the flying mile of 150.533. The sand was the reason for the car's lower speed. What did this have to do with the standard car line? Performance centered around the 265ci V-8, and the record set by the Corvette could only help to promote the standard passenger car line. It all pointed to the fact that Chevro-

let was a winner—that The Hot One was even hotter for 1956.

Chevrolet did well in NASCAR events as well. Chevrolet won the Convertible Division with Bob Welborn behind the wheel, and Jim Reed won the Short Track Division. Magazines and newspapers touted for the victories achieved by the 225hp small-block V-8. These race wins were all the more impressive when one considers the competition the bow ties boys faced. Ford certainly saw the Chevy threat and mounted an equally aggressive campaign, boosting its new 312ci Y-block V-

8 to 225hp. While Chevrolet seized the Convertible and Short Track Division titles in 1956, the Grand National title eluded them as the small-block V-8-powered bow ties were no match for the Chrysler 300s and their 354ci hemi-head V-8s.

Despite impressive racing wins, sagging sales were not bolstered. The recession in 1956 depressed automotive sales across the board. Chevrolet sales were down more than 11 percent in 1956, and total GM sales were down a staggering 23 percent. Ford sales were down even fur-

This was the engine that made The Hot One even hotter. The $242.10 RPO 411 sported dual inline Carter WCFB four-barrel carburetors, mechanical camshaft, 9.25:1 compression ratio, and dual exhaust. It was rated at 225hp at 5200rpm.

This Bel Air, owned by Jerome Cain, Fred Gaugh, and Terry Sheafer, is painted in India Ivory and Sierra Gold.

The '56 Chevrolet

It looks high priced—but it's the new Chevrolet "Two-Ten" 4-Door Sedan.

For sooner and safer arrivals!

It's so nimble and quick on the road . . .

Of course, you don't have to have an urgent errand and a motorcycle escort to make use of Chevrolet's quick and nimble ways. Wherever you go, the going's sweeter and safer in a Chevy.

Power's part of the reason. Chevrolet's horsepower ranges up to 205. And these numbers add up to *action*—second-saving acceleration for safer passing . . . rapid-fire reflexes that help you avoid trouble before it happens!

True, lots of cars are high powered today, but the difference is in the way Chevrolet *handles* its power. It's rock-steady on the road . . . clings to curves like part of the pavement. That's *stability*—and it helps make Chevrolet one of the few great road cars!

Highway-test one, soon. Your Chevrolet dealer will be happy to arrange it. . . . Chevrolet Division of General Motors, Detroit 2, Mich.

THE HOT ONE'S EVEN HOTTER

CHEVROLET

Traffic-test it— it's a beautiful thing to handle!

The Bel Air did not get all the advertising ink; there were after all, the Two-Ten and One-Fifty models. A V-8-powered Two-Ten is featured in this ad, as denoted by the large, shallow "V" underneath the hood badge. This car is shown in Crocus Yellow and Laurel Green.

ther, at more than 25 percent. Nevertheless, the 1956 Chevrolet received accolades from the automotive press for its 1956 line of cars, particularly the ones with the hottest V-8s.

The Automotive Press Responds

How did The Hot One fare with the automotive press for 1956? *Motor Life* editors wrote of the new 225hp V-8, "The new powerpack [sic] was noticeably livelier than last year's 180-bhp job at turnpike speeds . . . where the new car should make the biggest showing, 50–80 mph, time was lowered 3.5 seconds from stock '55 time, 0.9-second from the '55 powerpack's 12.9-second time." The editors also tested the 170hp version coupled to the two-speed Powerglide transmission. The editors recorded a 0–60mph time of 11.9sec with a top speed of 98mph.

The four-barrel, 205hp Super Turbo-Fire V-8 Chevrolet recorded a 0–60mph time of 8.9sec, with a 108.7mph top speed. *Motor Life* editors wrote the 1956 Chevrolet offered the ". . . best performance per dollar."

Tom McCahill, writing for *Mechanix Illustrated* in his usual florid style, said: "Chevrolet has come up with a poor man's Ferrari . . . here's an engine that can wind up tighter than the E string on an East Laplander's mandolin . . . well beyond 6000 rpm without blowing up like a pigeon egg in a shotgun barrel."

Road & Track editors got their hands on a two-door Two-Ten model with the 205 hp V-8 and manual shift. They achieved a credible 0–60mph time of 9sec, covered the quarter-mile in 16.6sec doing 80mph through the lights, and achieved a 111mph top speed. The editors extolled the engine, writing: "Without a doubt, the greatest charm of this car is its smooth, quiet-running engine. Even though the compression ratio is extremely high," the editors wrote, it was ". . . impossible to make it 'ping' on full throttle at any speed . . . the surge of power (actually torque) is there at all times, and knowing the ultra-short stroke, one gets the impression that this engine would be impossible to 'blow up' even under brutal treatment."

The 1956 Chevrolet's ride and handling were also duly noted. In its *1956 Features* book distributed to dealers, Chevrolet pointed out a number of refinements that had been made to the chassis and suspension. Front spring rates were reduced on all models except the convertible and the station wagons for a softer ride. The nine-passenger station wagons, in fact, received new six-leaf springs with a rate of 165lb per inch. To improve directional stability and handling, steering caster

Chassis for the 1956 Chevrolet were welded together, painted, then moved to the next chassis assembly phase which included mounting suspension components.

These assembly line workers used this two-position compression machine to collapse the coil springs so that the front shock absorbers could be mounted. While they performed this work, the rear leaf springs and shock absorbers were installed on the chassis.

The interior of this Bel Air was complemented with Copper Vinyl and Tan Pattern Cloth. The padded dash was option RPO 427. The three-speed column-shifted manual was the most effective means of extracting power from the high-revving 225hp V-8.

was increased one degree. Rear suspension durability was improved by widening the rear leaf spring hanger to three inches, up from two inches in 1955. Front wheel bearing durability was improved by using higher-capacity outer bearings. Rear axle bearings were now lifetime-lubricated and sealed at the factory. A redesigned fuel tank for the V-8-equipped wagons permitted a dual exhaust system.

Automotive journalists quickly discerned these refinements in the car's ride and handling. *Motor Trend* editors wrote of the ". . . absence of nose dive under all stopping conditions, including panic stops." Of the car's handling, the editors wrote, "Not only do we admire the steering ease . . .but believe you will be surprised to find that power steering isn't as noticeable as you might think. Not many things can upset Chevy's composure on the road; it weathers normal rigors with ease. Only when it's bounced hard by a bump, or rocked into a chuckhole in the midst of a fast turn does it betray its relatively light weight and semi-stiff suspension and skip from its initial track. Recovery from bumps, dips, and potholes is rapid, non-jarring in most cases, and free from wallowing or pitching."

Chevrolet engineers endeavored to get the best balance of ride and handling for the 1956 car

The Bel Air convertible was what top-down driving was all about in 1956. Almost 42,000 models were sold that year. This example is owned by Walter Cutlip of Longwood, Florida.

Chevrolet had six station wagons available for 1956. There was the TwoTen two-door Handyman, the two-door Nomad, the Two-Ten four-door Townsman, the One-Fifty two-door Handyman, the Two-Ten four-door Beauville, and the four-door Bel-Air Beauville.

line. *Motor Trend* editors wrote: ". . . passengers aren't pitched or rocked from side to side on twisting roads, or see-sawed back and forth in stop-and-go driving. Seats aren't soft either, but they soak up a great deal of chassis movement, [and] level out most minor disturbances."

Judging from these remarks, Chevrolet engineers achieved their goal. Other editors weren't always as positive, but for the vast majority of Chevrolet buyers, the ride and handling of the 1956 car line was perfectly adequate and downright pleasant.

Numbers Tell the Story

Prices did increase on the 1956 models, but it is interesting to compare prices not only to 1955 but to Chevy prices from the late forties and early fifties. The value leader was the One-Fifty Series, and the least expensive model was the three-passenger, two-door utility sedan at $1,734 with the standard "Blue-Flame 140" inline six-cylinder engine. This was an increase of $141 over the 1955 model. But the price on the aging 1949 two-door, three passenger Business Coupe had been $1,329. By 1952, the price on this model had risen to $1,524.

In the Two-Ten series, for example, the four-door, nine-passenger station wagon listed at $2,348. Nearly ten years before, the comparable 1947 Fleetline four-door, eight-passenger station wagon had a factory price of $1,893. The four-door, six-passenger sedan for 1956 had a price of $1,955—a sharp drop from $1,810 the previous year—while the price for the 1947 Fleetmaster four-door, six-passenger Sport Sedan was $1,345.

In the 1956 Bel Air line, the two-door convertible had a price of $2,344, an increase of $138 over 1955. However, the 1954 Bel Air convertible listed at $2,185, and the 1952 model had a price of $2,113. The four-door sedan in '56 listed at $2,068, a $136 increase from the previous year. The 1949 four-door Fleetline sedan had a price of $1,539. It should also be noted that the six-cylinder engine of the late forties and early fifties had only 90–115hp up through 1954.

The 1956 six-cylinder models had 140hp. Clearly, the 1956 Chevrolet car buyer was getting up-to-date styling, the latest engine technology with more power, and vastly improved ride and handling for a modest increase in price over the previous generation models.

Beyond price, the production numbers also told a story. Even with the addition of the four-door Sport Sedan, overall sales of the Bel Air line dropped nearly 100,000 units in 1956, compared to 1955. The Bel Air four-door sedan, by far the most popular model in the Bel Air line, dropped from 345,372 units in 1955 to 269,798 units in 1956.

The next most popular model Bel Air, the two-door hardtop Sport Coupe, also dropped in sales—from 185,562 units in 1955 to 128,382 units in 1956. The four-door Bel Air station wagon dropped from 24,313 in 1955 to 13,268 in 1956. The other Bel Air models also suffered sales losses. Nomad sales dropped from 8,530 in 1955 to 8,103 in 1956, but this was due primarily to having a full model year for sales. The new Bel Air four-door Sport Sedan was a bright spot, with 103,602 units sold, and this helped to staunch some of the sagging sales.

The recession was clearly having a severe effect on the automotive market.

Overall, Two-Ten model sales also dropped, from 806,309 units in 1955 to 770,955 units in 1956. The One-Fifty series suffered a similar percentage drop. Despite these disappointing losses, Chevrolet managed to maintain its sales lead over Ford, seizing 27.94 percent of the automotive market, compared to Ford's 23.68 percent.

There was one ray of sunshine that appeared in 1956 which had everyone at Chevrolet excited. That was the near completion of the new Chevrolet Engineering Center in Warren—a major resource that would help Chevrolet revolutionize its car and truck lines.

The Chevrolet Engineering Center

In December 1956, the new, long-awaited Chevrolet Engineering Center was unveiled to the automotive press. The site was four miles north of Detroit in the city of Warren, located just off Van Dyke Boulevard. Situated on fifty-four acres, the modern facility took three years to build. It was, in the words of Harry Barr, Chevrolet's chief engineer, the ". . . answer to an engineer's dream."

The Center consisted of three interconnected structures: a four-story office building, a one-story experimental shop, and a test laboratory. The shop, with presses ranging up to 2,500 tons, along with machine, paint, plastic, model-making, and sheet metal equipment, was now capable of turning out experimental vehicles ranging from futuristic show cars to pilot models that were carefully assembled prior to volume production.

In a news release at the time, Barr stated, "Working together at this facility as a closely co-ordinated team are all the diverse talents required to translate a new car from mind to metal. Here, piece by piece, the engines, transmissions, frames, and other vital components of the cars and trucks of tomorrow will be conceived, designed, built for prototype models, and subjected to exhaustive tests to prove their safety, performance and durability."

Louis Cuttitta vividly remembers one story involving Ed Cole and the respect all those who worked in the new engineering center had for him: "On the dynamometer cell doors where the engines were developed, it was called 'The Home of the Hot Ones.' It was a long hall with all these dyno cells. When Ed Cole walked down that hall, the dyno operators, technicians, and engineers, would clap. That's what you call respect."

The Chevrolet Engineering Code

Chevrolet wrote of its new Engineering Center in *Finger-Tip Facts*, and explained its engineering code, or credo. It was Chevrolet's engineering philosophy, if you will.

"Our first responsibility is to the customer. This explains our maxim 'Progress through constant improvement.' By progress, we do not mean that we grasp at every new or radical idea, but on the contrary, that we advance by means of fundamental improvements in design. We do not believe in trying our new ideas on the customer. Instead, we adhere to the policy of providing the customer with only the best of designs that we have proved to be sound through exhaustive tests. To give Chevrolet customers even better values, we do everything in our power to protect Chevrolet's position in the low-price field. Through unbiased tests of competitors' products, we keep posted on their designs, so that we know how to make ours even better. Our favorite policy is self-criticism. Before a new model goes into production, we do everything we can to find fault with it . . . so the customer will find it faultless. Above all, at no time do we compromise with safety in the Chevrolet!"

1956—A Great Year

The 1956 Chevrolet built upon the excellent foundation laid down by the 1955 model. It was a worthy year of refinement and improvement. For some, it was a better-looking car than the 1955, and owner surveys at the time confirmed this was one of the main reasons for buying their '56. Certainly, Chevrolet Chief Engineer Ed Cole agreed that the 1956 Chevrolets were even better than the '55s. In fact, the 1956 Chevrolets more than hold their own against some of the makes and models that came from other manufacturers that year. No wonder the car is considered classic!

This page and next page
Despite better looks, a full production
year, and exposure in the automotive
press, only 8,103 Nomads were sold
in 1956, even fewer than the previous
year's production of 8,530. This
example is owned by Steve and Chris
Mechling of Champaign, Illinois.

Smooth as Quicksilver

Perhaps no car is more identified with its production year than the 1957 Chevrolet. Show someone a 1957 Ford or Chrysler, and they will have to guess the car's vintage. The 1957 Chevrolet suffers no such identity crisis. That is due, in part, to the car's popularity today and its constant exposure in movies, magazines, and television. There is no mistaking that distinctive front end and those fabulous fins!

Today, car manufacturers could ill-afford an annual model change, but in the fifties it was a way of life for the manufacturer and the buyer. When one looks at the 1955, 1956, and 1957 Chevrolets side by side, they are dramatically different in appearance, but they are very similar under the skin. The 1957 model was the second facelift for the car. The 1958 Chevrolet would go through a complete evolutionary change yet again. Thus, the classic tri-Chevys are indeed a group unto themselves.

"SWEET, SMOOTH, and SASSY" was how Chevrolet described its cars in a sales brochure. "For '57 Chevy goes 'em all one better. New styling! New power—offering fuel injection for the first time! New interior luxury! Revolutionary new Turboglide—the newest, smoothest thing in automotive drives—and famous Powerglide, proved by billions of miles! These and many more advanced and exclusive new features mark Chevrolet the importantly new car in its field!"

The 1957 Chevrolet was built against a backdrop of national and geo-political events.

Dwight D. Eisenhower, comfortably into his second presidential term, issued the Eisenhower Doctrine in January, which called for aid to Mideast countries that resisted armed aggression from Communist-controlled nations. Later that month, the first Thor intermediate-range ballistic missile (IRBM), built by the Douglas Aircraft Company for the Air Force, was launched from Cape Canaveral, Florida; it exploded seconds after liftoff when a fuel tank ruptured. The Army's IRBM, the Jupiter, was first launched on May 31. Fewer than two weeks later, the first launch of the Atlas long-range ballistic missile took place at Cape Canaveral. It was the Soviets, however, who launched the first satellite, *Sputnik*, and America suffered an international black eye. The space race was on.

When we think of the fifties, many of our memories revolve around popular songs, movies and Broadway shows. Among the Top Ten tunes heard from car radios from coast to coast in 1957 were "Tammy" by Debbie Reynolds, "Love Letters in the Sand" by Pat Boone, "Chances Are" by Johnny Mathis, and "Bye Bye Love" by The Everly Brothers. *The Music Man* opened on Broadway, starring Robert Preston. *My Fair Lady*, which opened in 1956 and starred Rex Harrison and Julie Andrews, continued to pack them into the Mark Hellinger Theater and would become the longest

Imperial Ivory with Dusk Pearl was one of the new two-tone color combinations available for 1957. Chevrolet loaded up the Bel Air that year with as much chrome trim as the car could tastefully handle.

"Sweet, smooth and sassy" was the 1957 ad theme, though it did not appear in every ad for the 1957 Chevrolet line.

Below
One of the innovative improvements made in the 1957 Chevrolet regarded fresh air ventilation. Fresh air intake was moved from the base of the windshield to above the headlights. The new ventilation system provided a 22 percent increase in air flow over the old system.

'57 CHEVROLET! SWEET, SMOOTH AND SASSY!

Chevy goes 'em all one better for '57 with a daring new departure in design (looks longer and lower, and it is!), exclusive new Triple-Turbine Turboglide automatic drive, a new V8 and a bumper crop of new ideas including fuel injection!

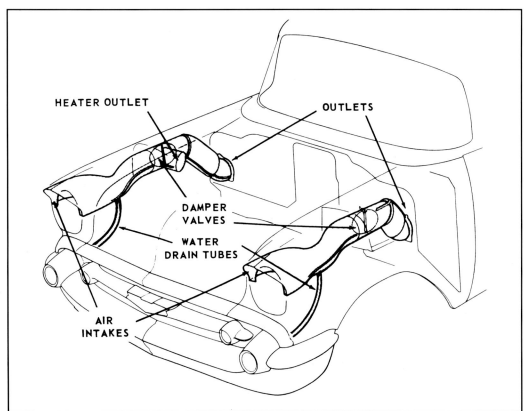

HEATER OUTLET

OUTLETS

DAMPER VALVES

WATER DRAIN TUBES

AIR INTAKES

running Broadway show of the decade. Art Linkletter was a household name with his TV show "Kids Say the Darndest Things" and his best-selling book by the same name. Brigitte Bardot set the imaginations of countless men aflame in *And God Created Woman*, except in Philadelphia, Fort Worth, Abilene, Memphis, and Providence, where it was banned. Oscars were awarded to *The Bridge Over the River Kwai* for Best Picture, to Joanne Woodward for Best Actress for her portrayal in The *Three Faces of Eve*, and to Alec Guinness for his role in *The Bridge Over the River Kwai*.

America, especially young boys, had a fascination with jets, not just rockets and cars. Numerous headline-making records were set by American jets in 1957. Maj. Gen. Archie J. Olds, Jr., leading three Air Force B-52 bombers, flew the first non-stop, around the world flight in 45 hours, 19 minutes while averaging 525mph. Air Force Major A.E. Drew flew an F-101 Voodoo and set a record for jet aircraft by flying 1,207.6mph. Commercial jets were flying in England and would soon start service in the United States. Perhaps all this emphasis on speed during the mid-fifties had a bearing on the look of the 1957 Chevrolet.

Launching the Fin Era—and More

The 1957 Chevrolet actually started taking shape in Chevrolet's styling studios in 1954. The Division had a significant budget to change the car's looks, but the doors, roof, and trunk lid had to be retained. Still, that allowed a fair amount of design flexibility; the entire front end, including the hood and the rear fenders could be altered.

Styling exercises at the studio in 1954 reflected a planned reduction of the cowl height by 1 1/2in as a result of a new fresh air intake. Instead of drawing fresh air from the high pressure area at the base of the windshield and passing it

A truly glorious sight: 1957 Chevys as far as the eye can see. One of the last pieces of trim to be mounted were the headlight bezels with air intakes.

through a large plenum, air would be picked up through screened areas above the headlights and fed through ductwork directly into the firewall and from there to the interior vents. Doing away with the plenum permitted a lower cowl, increased visibility from the larger windshield, and a lower, flatter hood.

The Chevrolet studio searched for a new front end theme. Some clay studies employed the existing front bumper with a considerably different grille/headlight configuration. One version showed the headlights above and the parking lights below in a stacked configuration, with a concave grille featuring nine vertical, concave chrome bars. The top of the front fenders displayed chrome windsplits and the hood had a redesigned ornament. Another clay version sur-

rounded the grille area with a massive lower bumper and a lighter upper bumper. This approach progressed in 1955. The final front end designed that emerged—very Oldsmobilesque in appearance—was best described in the *1957 Chevrolet Passenger Car Features* book the Division published to describe changes and improvements in its cars:

"The functional and decorative grille-bumper combination is composed of three main elements which blend into the sheet metal contours and contribute to the low, clean-cut lines of the front end styling. The bright metal header bar, attached to the hood, arches to meet the bumper proper. The bumper is a massive element which replaces the lower portions of the fender corners and sweeps under and across the vehicle width. At ei-

ther end the contour is extended forward, forming massive decorative elements to replace the former applied bumper guards. The grille design features a horizontal center bar, terminating at either end in circular parking lights and, at its center, contains the Chevrolet medallion. The center bar is located on a lattice pattern screen of long horizontal rectangles which fills in the grille opening. The screen is of gold anodized aluminum on Bel Air models and bright anodized aluminum on Series 2100 and 1500."

With the Chevrolet medallion now in the grille, the forward edge of the hood featured the word "Chevrolet" above a shallow "V" on V-8-equipped models; six-cylinder models displayed only the Chevrolet name in larger script. On top of the hood, the winged hood ornament on the 1956 model gave way to two chrome windsplits, with the circular contours in the sheet metal running from the windsplits almost to the back of the

Here, the floorpan is welded in place. Note the welded seam on the right rear fender. This would be filled with lead by the worker in the background on the left.

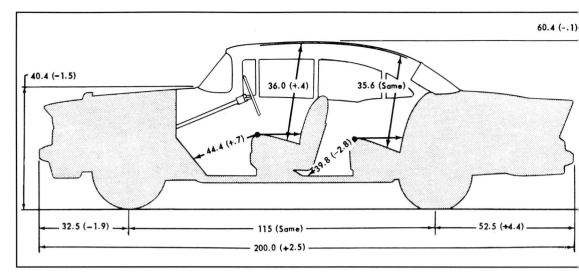

60.4 (−.1)

40.4 (−1.5)

36.0 (+.4)   35.6 (Same)

44.4 (+.7)

39.8 (−2.8)

32.5 (−1.9)   115 (Same)   52.5 (+4.4)

200.0 (+2.5)

Above
This illustration shows the dimensional differences between the 1957 and 1956 Chevrolets. Front overhang decreased nearly 2in while rear overhang increased more than 4in. Overall length increased 2-1/2in.

Left
Final inspection of bodies just prior to their moving to the overhead line for mating to the chassis. Here, a worker inspects a One-Fifty Handyman wagon.

hood. The headlights were now prominently hooded, encircled with bright chrome trim, and spaced 1.3in farther apart. Three D-shaped indentations were added to the sides of the front fenders near the headlights; on the Bel Air these received gold anodized aluminum inserts.

The development of full-blown fins on the 1957 Chevrolet were only mildly hinted at in the 1956 model. It was quite a stylistic departure, but it turned out to be in step with the times, as all the manufacturers were embracing the "fin mentality." The intent was to make the car look longer. In fact, the car was longer as a result of the restyling effort. Front overhang decreased 1.9in, but rear overhang increased 4.4in. Overall length grew from 197.5in for 1956 to 200in in 1957 for the standard cars; station wagons, however, dropped in length, from 200.8in in 1956 to 200in to match

This was the quintessential Nomad— the 1957. The car took to the restyling effort very well. Chevrolet moved to an open grille, massive bumper, and more chrome just about everywhere. Erol and Susan Tuzcu of Del Ray Beach, Florida, own this fine example.

A Two-Ten Club Coupe body begins the mating process with its chassis. Note the two large ventilation air intakes in the firewall. A hand-written note on the windshield orders the entire car to be polished at the end of the assembly line.

the sedans, coupes, and convertible. Overall height for all the models remained unchanged. Overall width was trimmed 0.4in for all models.

Brightwork trim molding was again used along the sides of the car to distinguish the three models. The One-Fifty used a chrome spear running from the middle of the fin at the rear to just before the front doors on the four-door models and just beyond the front doors on the two-door models. The louvered, diagonal piece of brightwork from the spear up to the window was identical to the previous year. With the front fenders now barren of brightwork, the name "Chevrolet" appeared there in chrome. On the Two-Ten and Bel Air models, the design and application of brightwork molding along the sides was identical. A spear-type molding began immediately to the rear of the headlight and arched gently downward along the full length of the car to blend into the

rear bumper. At the rear quarter area, another molding branched upward, then swept horizontally to the rear, terminating at the rear fender crown molding. On the Two-Ten model, the word "Chevrolet" appeared between the two pieces of molding and near the fender crown molding. On Two-Ten models with two-tone paint schemes, this area between the moldings received the contrasting paint color. On the Bel Air, this area between the moldings received a rolled, anodized aluminum panel with shallow, horizontal flutes. The word "Bel Air" and the Chevrolet medallion appeared on this panel.

The rear bumper was dominated by the bumper guard and taillight combination which consisted of three visual elements. The upper crescent portion contained the taillight, stop light, and turn signal lights. The center oval section had a stamped, anodized aluminum cover plate that

Here, the finished bodies receive the front and rear seats after the interior panels and headliner have been installed. Two-tone paint was a popular choice with 1957 car buyers; all the cars in this picture except the one to the right have two-tone paint schemes.

was replaced when the accessory back-up lights were ordered. Below this was a mirror-image crescent shape that originally was to function as the exhaust outlet, but was dropped when a similar attempt on Cadillacs resulted in sooty bumpers. On the Chevrolets, this crescent-shape area remained solid and was blacked-out. All Chevrolets received chrome crown molding along the trailing edge of the fender and along the top of the fin. On the One-Fifty and Two-Ten models, the chrome crown molding on top of the fender extended forward roughly 12in from the tip of the fender. On the Bel Air models, this molding extended forward almost 4ft. The left fin contained the fuel filler cap

above the taillight assembly; the chrome molding was hinged for fuel filler cap access.

Interior stylists of the 1957 Chevrolet abandoned the basic design used for the instrument panel the previous two years. The *1957 Chevrolet Passenger Car Features* book described the new panel this way: "Instruments are installed under a deep, flat hood, superimposed on the instrument panel crown, directly in front of the driver. The hood is framed in chrome on Series 2400 (Bel Air) and 2100 (Two-Ten) models. Full-faced gauges, with red indicators and tell-tale lights for the generator and oil pressure, permit the driver to take readings at a glance. The large, round speedome-

ter, containing the odometer and high-beam indicator as well as the automatic transmission quadrant, dominates the cluster. At the left of the speedometer is the temperature gauge, and at the right the fuel gauge. The radio speaker grille is relocated to the top of the instrument panel crown."

All the control knobs and ventilation controls were redesigned and repositioned. The door-to-door instrument panel appliqué was the same shape but had a new texture pattern. Of course, the interior and seating areas of all three models were redesigned, boasting new color combinations, fabric designs, and vinyl textures.

There was some reshuffling of the solid and two-tone exterior colors for 1957. There were now sixteen solid colors: Onyx Black, Inca Silver, Imperial Ivory, Matador Red, Harbor Blue, Larspur Blue, Tropical Turquoise, Surf Green, Highland Green, Colonial Cream, Coronado Yellow, Canyon Coral, Sierra Gold, Adobe Beige, Dusk Pearl, and Laurel Green. The two-tone combinations were Onyx Black and India Ivory, Imperial Ivory, and Inca Silver, Larspur Blue and Harbor Blue, India Ivory and Larkspur Blue, India Ivory and Tropical Turquoise, Surf Green and Highland Green, India Ivory and Surf Green, India Ivory and Coronado Yellow, Onyx Black and Colonial Cream, India Ivory and Colonial Cream, India Ivory and Canyon Coral, Adobe Beige and Sierra Gold, India Ivory and Matador Red, Colonial Cream and Laurel

Green, and Imperial Ivory and Dusk Pearl.

The 283ci V-8

Chevrolet boasted a number of engineering firsts for 1957. Topping the list was the larger-displacement 283ci V-8. The 265ci V-8 remained in the engine lineup, but the new 283ci V-8 understandably stole all the thunder. Was an increase of less than 20ci really worth the effort?

"Car design and engine design don't just happen, they evolve," Fred Sherman, a dyno technician who worked on the small-block V-8 at the time said. "After running the 265-cubic-inch V-8 for two years, you know, invariably, no one's going to be satisfied, because to become satisfied is to become complacent. Dissatisfaction is one of the most necessary ingredients there is to move ahead. Well, the question was, What if? Suppose we open

the bore. What could we look for? Volumetric efficiency, more breathing, and more power from greater displacement. We'd go through a complete series of tests. There'd be a heat rejection test, there'd be testing of different cam profiles, there were different diameter valves to be used—all with one thing in mind: Let it breath better."

For 1957, prospective buyers had a choice of three engine displacements: the 235ci "Blue Flame" six-cylinder, one version of the 265ci V-8, and an array of 283ci V-8s. Carbureted versions of the 283 included a two-barrel V-8 with 8.5:1 compression, 185hp at 4600rpm, and 275lb-ft of torque at 2400rpm; a four-barrel version with 9.5:1 compression, 220hp at 4800rpm, and 300lb-ft of torque at 3000rpm. The dual four-barrel 283 with 9.5:1 compression was rated at 245hp at 5000 rpm with 300lb-ft of torque at

The floorpan takes shape. Here, additional sheet metal parts are welded to the floorpan.

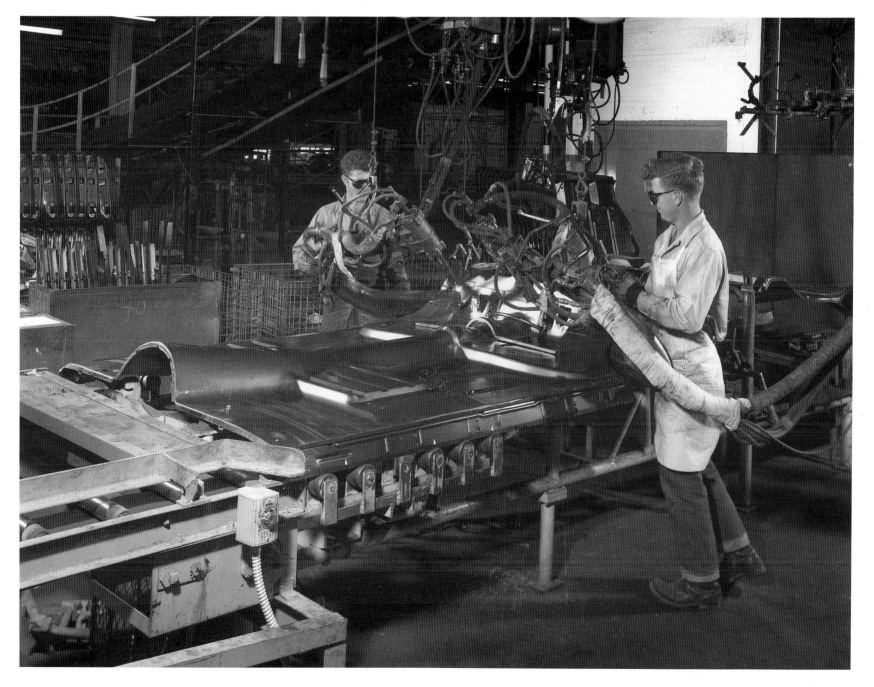

The cylinder heads were redesigned for 1957, with intake and exhaust ports enlarged for improved breathing. The cylinder block was also redesigned to reduce the possibility of cylinder bore wall distortion from the cylinder head bolts.

3800rpm. There was also a competition version of this last engine which included mechanical lifters and 9.5:1 compression rated at 270hp at 6000-rpm with 285lb-ft of torque at 4200rpm. Chevrolet planned to introduce fuel-injection on the 283 later in the model year. One version of the injected 283 would develop 250hp at 5000rpm with 9.5:1 compression, and the highest output version, with mechanical lifters and 10.5:1 compression, would develop 283hp at 6200rpm with 290lb-ft of torque at 4400rpm.

The greater displacement was achieved by increasing the bore 0.125in, making it 3.875in. The 283 retained a main bearing journal diameter of 2.3in, but front and intermediate main bearings were 0.0625in wider. Rod bearing journal diameter remained at 2in.

But there was much more to this new small-block than just additional displacement. Chevrolet wanted more than just better breathing and more power; it wanted more accurately timed and more uniform mixture burning, improved lubrica-

tion, and greater overall durability. To achieve this, cylinder head intake and exhaust ports were enlarged and coupled with the new "ram's horn" exhaust manifolds. Except for the competition versions, the 283 had a high-lift hydraulic camshaft—0.3987in compared to 0.3336in for the 265.

A new engine lubrication system was also designed into the small-block. Previously, a milled slot on the outer diameter of the camshaft's rear journal metered the amount of oil and reduced the pressure of oil which passed from the main oil gallery to the lifter galleries via two drilled holes in the camshaft's rear bearings. In the new lubrication system, oil was fed to the lifter galleries from the main gallery under direct pressure, and the milled slot that had metered the oil was deleted.

Detail improvements included a redesigned second compression piston ring to improve compression and oil control under certain operating conditions. An improved oil baffle covering the surface of the inlet manifold exhaust crossover passage minimized the possibility of oil carboniza-

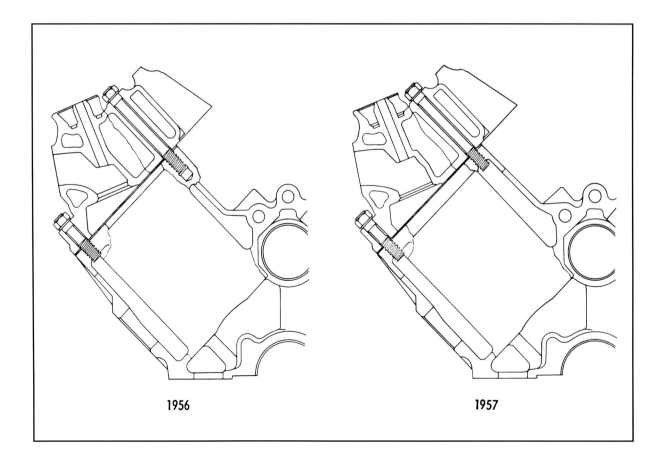

1956   1957

tion in the valve lifter areas. A new distributor used on all 283s except the competition versions permitted accurate point settings to be made while the engine was running. An improved carburetor choke design resulted in smoother operation and increased fuel economy during cold starts. The spark plugs were designed so that the electrodes reached farther into the combustion chamber for shorter flame travel and more uniform combustion mixture. At the other end of the spark plugs, a metal deflection shield protected the spark plug boot and wire from exhaust manifold heat. Dual exhaust systems were used on all 283s, but there was now a crossover pipe which equalized left and right bank cylinder pressures, quieting exhaust and improving torque slightly.

The new 283 with all its improvements was certainly of interest to bow tie buffs and prospective buyers, but it was the much-rumored new fuel injection system that was most eagerly awaited by Chevy enthusiasts, racers, and automotive magazine editors.

Ramjet Fuel Injection

The new fuel injection system Chevrolet introduced in 1957 was designed by the Rochester Products Division of GM, under the direction of E. A. Kehoe and Donald Stoltzman, working with John Dolza of GM's engineering staff, and Cole, Barr, and Arkus-Duntov at Chevrolet. It was a continuous-flow mechanical unit which had flexibility in calibration and adjustment. It combined two signals—mass air flow and manifold vacuum—to correctly meter the precise air/fuel ratio according to the engine's needs. Louis Cuttitta, who had distinguished himself as an expert in carburetion, was invited to work on the production version of GM's fuel injection system.

Ramjet Fuel Injection appeared on the 283ci V-8 in 1957. Chevrolet boasted one horsepower per cubic inch from this engine. It was available in both the standard car line and the Corvette, but this option proved very hard to get due to limited production.

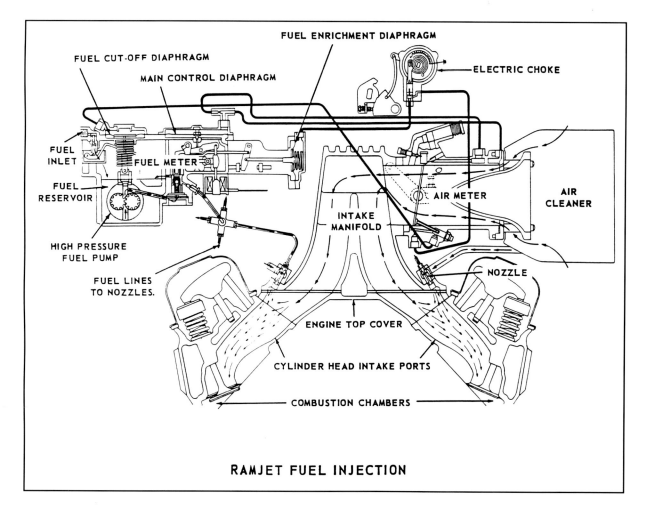

RAMJET FUEL INJECTION

The hidden fuel filler cap feature for 1957 was modified to conform to the car's new styling.

"I got an invitation to go to engineering staff and listen to a guy by the name of John Dolza," Cuttitta remembered. "He was the inventor of the early 'A' fuel injection system for General Motors. I heard him speak and looked at this fuel injection system—which was really a pressurized carburetor—and I got interested in it. He had already heard who I was, since I was instrumental in working with Cole on carburetors, and I got to do the development work on the Rochester 'A' fuel injection system. Ironically, at this time, there was an old buddy of mine who had worked at Continental Aviation who now worked at Rochester Products—Frank Sceabica, a development engineer there for the 'A' system. Between the two of us, we had the world by the tail. We got in touch with Fred Frincke, Cal Wade, and Denny Davis, who actually did the manifold, putting lines down on paper and getting the parts made. That's how, by

early 1956, we had the first fuel injection system on the Corvette that ran at Sebring and then went on to the Nassau race. And, of course, Duntov oversaw the whole program."

The *1957 Chevrolet Passenger Car Features* described in detail the advantages of Ramjet Fuel Injection over carburetion: "Carburetors and their various disadvantages are eliminated. For instance, fuel atomized by air passing through the carburetor venturi is carried all the way from the carburetor, through the intake manifold, to the intake ports at each cylinder. This makes it necessary to provide intake manifold passages large enough for easy breathing, but small enough to maintain adequate air flow velocity at idle so that the fuel will not settle out and upset the proper mixture ratio. Furthermore, carburetors must be heated to help vaporize the gasoline to prevent the formation of ice in the carburetor air-fuel passage on cold starts due to the refrigerating action of the carburetor venturi and the atomization of the fuel. The heat required expands the incoming air, making it less dense, causing loss of power.

"On the other hand, the Chevrolet Ramjet Fuel Injection system supplies fuel under pressure right up to the cylinder head intake ports through individual fuel lines. Separate passages are also provided to supply clean air, unencumbered by gasoline vapors, all the way to each individual cylinder head intake port. Here the fuel, finely atomized by the injection nozzles, mixes thoroughly with the fast-moving air and is drawn into each combustion chamber, thus providing precise and uniform cylinder-to-cylinder distribution. Because fuel is vaporized at the cylinder head intake ports, and the incoming air does not pass through a venturi of the carburetor type where refrigerating action takes place, the induction system does not require heat to prevent icing or help atomize the fuel. As a result, the incoming air is not subjected to external heat and a larger quantity of air is drawn into the combustion chamber. Volumetric efficiency is thus improved and engine power increased."

As Chevrolet ads would claim, the highest rated Ramjet Fuel Injection 283 V-8 had achieved the engineer's dream of 1hp per cubic inch. It was a very expensive option, costing $484.20. The 250hp version carried the same price.

Rochester Products Division officially announced Ramjet Fuel Injection in a press release dated Wednesday, October 17, 1956. However, despite all the claims made by Rochester regarding the availability of the option, it initially proved very hard to order one in your '57 Chevrolet. No one was more frustrated than magazine editors who had been clamoring for a car to test. In the February 1957 issue of *Motor Life*, the editor wrote, "The dramatic announcement of f.i. can be disregarded, for all practical purposes, where the average driver is concerned. The unit will be in limited supply; furthermore, the extra cost of the set-up makes carburetion still look pretty good." Racer Brown, in the March 1957 issue of *Hot Rod* magazine, had similar sentiments: "Chevrolet's first may have been premature but it sure stirred a storm of interest. However, interest is fickle in we mortals, and a first in anything isn't worth a plugged dime if it can't be bought . . . all the Rochester units are in the hands of the factory or the dealers, and none have sold to the public, except for a few for racing purposes."

Nevertheless, Chevrolet advertised its fuel-injected 283 V-8, more an effort of product promotion than availability. In the June 1957 issue of *Motor Trend*, the ad read in part: "In American automobile engineering, the magic milestone is this: one horsepower per every cubic inch of engine displacement! Naturally, we're proud. Because this is proof, in cold figures, of the extra efficiency of Chevrolet's advanced valve gear, free-breathing manifolding, and ultra-short stroke."

Just how many of these fuel-injected units found their way into Chevrolets in 1957? The National Corvette Restorers Society received documented information from Chevrolet that, of 6,338 Corvettes that year, 1,040 had Ramjet Fuel Injection, and roughly 1,500 of the other passenger cars came so equipped.

Beauty More Than Skin-Deep

Chevrolet boasted that the 1957 model was a "...product of extensive engineering development." Besides the scene-stealing 283ci V-8s, what else was really new under that 1957 skin?

The rolling chassis received a number of improvements and refinements. The frame's front cross-member-to-sidemember braces were redesigned to form a box section when welded to the frame, instead of nesting within the frame members; this made the front of the frame stronger and more rigid. Chevrolet engineers worked to improve the car's ride for 1957. Tires now were mounted to smaller 14in rims and were rated at 22psi, 2lb less than the previous year. The tires were a substantial 8in wider and were 1in smaller in diameter.

The lower control arms were spherical joint assemblies and were redesigned to improve front suspension durability. The socket retaining the ball stud was cast- instead of stamped-steel.

Vehicle handling was improved by changing the mounting of the rear springs to a more hori-

The vertical fin reached its zenith in the 1957 Chevrolet. The fins appeared exaggerated on the Nomad due to its forward-sloping tailgate. Sales of the Nomad dropped to 6,534 units for 1957.

zontal position to attain what Chevrolet called "zero steer" with a driver and front seat passenger. This was done by simply raising the front leaf spring hanger location 1/2in. For 1957, Chevrolet switched from negative camber to positive camber leaf springs to compensate for the small-diameter tires and retain proper vehicle height. Front and rear shock absorbers were redesigned and recalibrated to accommodate the changes.

The rear axle received improvements in the form of larger and higher-capacity rear wheel bearings. The bearings were now lubricated via the differential fluid; previously, they had been factory-sealed. The differential carrier and case were substantially strengthened by increased thickness and added reinforcing ribs. Chevrolet also changed from barrel-type roller bearings to tapered roller bearings where the axle entered the differential. The differential gears were now phosphate treated to increase their durability and aid in break-in.

Perhaps the biggest piece of engineering news for 1957, aside from the larger-displacement small-block V-8 and the availability of Ramjet Fuel Injection, was the introduction of the three-speed Turboglide automatic transmission. For many power train engineers at Chevrolet, the Turboglide

These two graphs show the power curves of the 220hp Super Turbo-Fire 283 and the 250hp 283ci Corvette V-8 with fuel injection. Performance had entered a golden age at Chevrolet and would continue until the advent of emission controls in the early seventies.

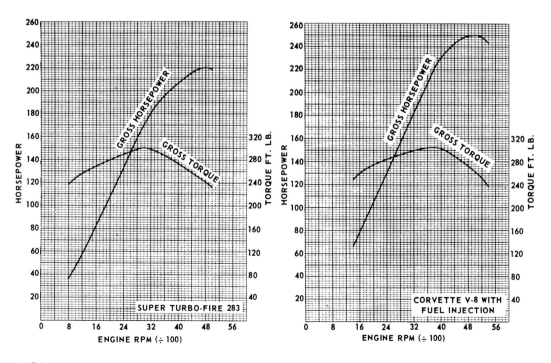

124

After the front end was bolted in place, these workers mounted the hood.

transmission was as significant an event as the small-block V-8 itself.

The goal of Chevrolet transmission engineers was to provide better torque multiplication by using three gears instead of two and to make shifts as seamless as possible. The Turboglide transmission was described in the *1957 Passenger Car Features* book this way: "Forming the basis for many of its functional features, Turboglide embodies a unique hydro-dynamic driving principle that eliminates a clutch or band-type engagement from standstill to top speed. Three turbines of a five-element torque converter are connected individually to the output shaft through the elements of two simple planetary gearsets. Thus, total torque multiplication is the product of both the torque con-

verter and gear ratios, combining liquid smoothness with the efficiency of gears. The broad ratio coverage is achieved through an ingenious coupling of the turbines to the planetary elements. It is an arrangement that permits a continuous amplification of the ratio change taking place in the converter."

The key to the Turboglide's smoothness was the use of three turbines and a variable pitch stator within the torque converter. The contribution of each turbine to the total driving torque was determined by vehicle speed and the power output of the engine. The range of the turbines was designed to overlap. The turbines were connected to the output shaft through the elements of the two planetary gearsets. The clutches worked in con-

The 283ci small-block V-8 was introduced in 1957. It proved to be an ideal blend of displacement, power, and fuel efficiency and remained in production for many years. The few changes made to the 283 after its introduction included a lower-profile air cleaner and a five-blade fan with viscous clutch.

This page and next page
The 1957 Nomad continued to have the same interior compartment flexibility as the more plebeian four-door station wagons in the Chevrolet line. Note the rear seatback retaining bar, which kept the seatback securely in either position.

junction with the planetary gearsets to aid in smooth application of torque to the driveshaft.

The gear selector sequence from left to right with the Turboglide was P for Park, R for Reverse, N for Neutral, D for Drive and HR for Hill Retarder. The HR position was designed to provide vehicle braking through the transmission while descending steep grades. This was particularly appreciated when towing a boat or trailer.

Ed Cole saw the possibility for tremendous weight savings by using aluminum for the Turboglide's transmission housing. When the Turboglide entered production, the transmission housing was the largest single die-cast aluminum part in the world. Yet, the transmission housing weighed only 15lb. The Turboglide automatic transmission option added only 4lb to the car's weight over that of the manual transmission. By comparison, the two-speed Powerglide transmission added 92lb to the car's weight.

The Turboglide automatic transmission was

less than $50 more than the Powerglide, which made it a bargain. However, new car buyers who were leery of the new Turboglide could opt for the proven Powerglide, which had also received improvements for 1957. Although the changes to the Powerglide were minute, they were important: Tip chamfer on the gear teeth was reduced to improve gear life, oil flow was improved to better cool the clutch faces, and primary washers were altered for increased durability.

Prices . . . and Extra-Cost Options

Car buyers today take for granted the many standard features new cars offer. In the fifties, one sure-fire way of making cars more affordable was to make many features optional or an accessory. Chevrolet was, after all, the "leader in the low-priced field." And Chevrolet worked to keep it that way. For 1957, there were again four models in the One-Fifty Series: the six-passenger four-door sedan ($2,048), the six-passenger two-door sedan ($1,996),

Front coil springs are installed then compressed to permit installation of the shock absorbers. In the background, a worker installs the rear leaf springs.

Next page
The Two-Ten series has often been overlooked by collectors and restorers, and One-Fifty series Chevrolets of the Tri-Chevy era are virtually impossible to find in stock condition. More than 162,000 Two-Ten series two-door sedans were built in 1957; this car is one of the few survivors today.

the three-passenger two-door utility sedan ($1,885), and the six-passenger two-door station wagon ($2,307).

The eight distinct models in the Two-Ten series, all six-passenger cars except for one of the station wagons, were carried over for 1957: the four-door sedan ($2,174), the four-door hardtop Sport Sedan ($2,270), the two-door sedan ($2,122), the two-door Del Ray Coupe ($2,162), the two-door hardtop Sport Coupe ($2,204), the four-door station wagon ($2,456), the nine-passenger four-door station wagon ($2,563), and the two-door station wagon ($2,402). The Bel Air line continued with seven models, all six-passenger except for the five-passenger convertible: the four-door sedan ($2,290), the four-door hardtop Sport Sedan ($2,270), the two-door sedan ($2,238), the two-door hardtop Sport Coupe ($2,299), the two-door convertible ($2,511), the four-door station wagon ($2,580), and the two-door Nomad ($2,757). The base V-8 added an additional $100 to each model.

As mentioned earlier, these cars were priced so reasonably because they were, essentially, stripped. How much more would the car cost with most of the items that are standard on today's Chevrolet? Chevrolet's most popular model for 1957 was the Two-Ten four-door sedan, with 260,401 units sold. The average buyer of one of these cars was, say, an executive with General Electric. He had been planning to buy a new Chevrolet for some time, and the new 1957 models convinced him this was the year to buy. He had recently received a raise, so he opted for the V-8 model listing at $2,274. The executive had a wife and two small children, and he probably wouldn't buy another new car for five years, so he carefully reviewed the options list with his wife. The 220hp Super Turbo-Fire four-barrel V-8 with dual exhaust was only $43.05, so he checked it off while his wife was pouring over the color chart. Living in a suburb of Boston, Massachusetts, they neither needed nor could afford the $430 air conditioner, but they most definitely selected the Airflow heater for $74.90. His wife insisted on the two-speed Powerglide transmission for $188.30 and power steering for $69.95. Electric windshield wipers were only $11.30, so they checked them off as well.

The dealer demonstrated the electric power windows on the car in the showroom. Electric windows were a real status symbol, but at $102.25 they had to really think about it. They finally laughed and said yes. White sidewall tires would really dress up the car, but they chose the less-expensive 7.50-14/4PR option at $31.60 instead of the 7.50-14/6s which cost $85.00. "Would you like a two-tone paint scheme on the new car," the salesman asked. It only costs 21.55.

"Yes," the wife said smiling, "I've already chosen the colors I want: India Ivory and Colonial Cream." The salesman assured them the Ivory vinyl and Charcoal Pattern Cloth interior would be an excellent compliment. A radio was a must. The push-button model seemed high at $89.50, so they opted for the manual-tuning model with antenna for $65.50. The salesman tried to steer them toward a long list of dealer-installed options but was politely turned down.

They asked the salesman to total the cost of the car and options. Double checking his figures, the final tally came to $2,882.40. With well-rehearsed ease, the young couple took on pained expressions and started to get up from their chairs. The salesman assured them they could probably strike a deal, and with their trade-in, it could be most affordable! After some haggling, they all agreed to a price, and the young couple left smiling.

The young executive's manager also set his sights on a new 1957 Chevrolet. However, he could afford to indulge himself. And he did.

He chose a Bel Air convertible. The base price of the six-cylinder Bel Air droptop was a reasonable $2,511, so he decided to order the car "loaded." He added the 283hp Ramjet Fuel Injection V-8 for $484, backed by the $231 Turboglide transmission. Being a hot rodder at heart, he also ordered the 4.11:1 limited-slip differential for $45. He added power steering for $69.95, power brakes for $37.70, electric windows for $102.25, and electric seat controls for $43.05. With all these power accessories, he thought it was wise to order the heavy-duty battery for a mere $7.55 and the 40-amp generator for $80.70. E-Z Eye tinted glass ($32.30) would reduce the sun's glare. Since he intended to enjoy top-down driving, he did not order the air conditioning, but the cold Boston winters made the $74.90 heater a mandatory option. He wanted to dress up his Bel Air convertible and he began with the $85 deluxe whitewall tires. The padded instrument panel ($16.50) seemed like a nice idea, so he ordered it, too..

The salesman, never one to overlook a substantial commission, informed the buyer of the many dealer-installed options. By far the most eye-catching was the Continental wheel carrier. It was an elegant touch and well worth the $129.50 suggested retail. At the other end of the car, the buyer asked for the front bumper guard cushions ($3.90). The Continental kit would force him to pull the car all the way into the garage, and the bumper cushions would prevent a jarring stop against the wall. The garage was a bit narrow, too, so he ordered door guards ($3.45)—a nice decorative touch as well. Since mirrors were not standard, he chose the inside rearview mirror ($4.95) and the driver's side outside body-mount unit ($4.40). The electric clock would be a very practical option ($17.85), and his wife would appreciate the vanity mirror ($1.60). The salesman informed him of the underhood lamp ($2.10), interior courtesy lamps ($2.75), glove compartment lamp ($1.15), luggage compartment lamp ($2.10), and backup lights ($12.00). The buyer nodded in agreement. To protect the beautiful carpeting, in went two deluxe floor mats ($6.95).

The salesman was warming up to this buyer, so he took him over to the Bel Air on the showroom floor and demonstrated the Wonder Bar radio and left rear antenna ($112.00). The buyer was all smiles, so the salesman suggested a matching dummy right rear antenna ($6.95) to give a sleek and balanced look. The salesman demonstrated the rear speaker ($12.95) which, working with the front dash speaker, gave the interior a rich, full sound.

When the salesman tried to push even more options, the buyer held up his hand and informed him that was what he wanted—now what was the total price? Adding up all the options took some time, and the grand total came to $4,139.50. The buyer sat there stunned, but tried not to reveal his shock. The options alone totaled more than $1500! Instantly, the salesman leapt into action. The buyer's 1952 Chevrolet could act as a trade-in, the dealer could probably make a deal to cut a

The Two-Ten was available in three vinyl and cloth color combinations. The Jacobsens' car is upholstered in Ivory Vinyl and Charcoal Pattern Cloth. Totally devoid of interior options or accessories, this car was purchased without so much as a radio.

Previous page
The Two-Ten series Chevrolet employed the same side chrome molding as the Bel Air, minus the stainless-steel appliqué. This Imperial Ivory car is owned by Bill and Barbara Jacobsen of Odessa, Florida.

What made this Two-Ten such a collectable car, and helped it avoid the ignoble fate of the crusher, was engine option no. RPO 411. The small-block V-8 displacement was increased from 265ci to 283ci for 1957.

couple of hundred dollars from the price of the car, and the cost of financing his new car through his bank was very reasonable with the interest rates so low. "And," the salesman added, "it will probably be the only Bel Air convertible like it in the entire country. It will turn heads wherever you drive it." They buyer beamed, agreed, and signed the order.

It is interesting that while cars had changed dramatically prior to the Tri-Chevys, and have changed even more since, the emotional aspect of buying a car has not changed at all. One of the activities surrounding cars—racing—has always had a direct or indirect impact on the way cars have

Option no. RPO 411 could be ordered with the high-lift, high-performance camshaft rated at 270hp at 6000rpm for $242.10 or with a more streetable camshaft offering 245hp at 5000rpm for $209.85.

Engineers had originally intended for the exhaust to exit through the opening beneath the bumper, but discoloration caused by heat made them change their minds. When optional backup lights were ordered, they filled this space.

Lloyd Brekke's '57 Bel Air Sport Coupe poses with a 1958 Beechcraft Bonanza. This was a popular model for Chevrolet that year, selling more than 168,000 units. Brekke's car was ordered with E-Z Eye Glass (RPO 398) and rubber front bumper cushions.

The 1957 Chevrolets received a redesigned fresh air ventilation system. A blower mounted behind the right inner fender drew air through vents placed above the headlight and passed the air through the firewall, where it was directed to vents in and below the dash.

been marketed. That approach has risen and fallen with the mood of the corporation, the country, and even the government. This means of marketing the Chevrolets had peaked and was falling, but that had no effect on the fact that the cars themselves were the most exciting yet.

The Hottest of the Hot Ones

Racing was conspicuously absent from Chevrolet's 1957 advertising. This was due primarily to efforts by the Automobile Manufacturers Association (AMA) to discourage car makers from engaging in track warfare. The AMA was fearful that this promotion of racing might incur the wrath of regulators in Washington, so racing-

oriented ads disappeared almost overnight. Fortunately, the performance image of street cars remained, but this aspect was de-emphasized in advertising. The following text from one of the 1957 Chevrolet ads was representative of the new tone: "Bring on the mountains! This new Chevy takes steep grades with such an easy-going stride you hardly even give them a thought. There's horsepower aplenty tucked away under that hood—the Chevy kind, just rarin' to handle any hill you aim it at."

Still, the push for greater V-8 performance was simply endemic to America's postwar mood. It was really the first performance surge since the thirties, when the likes of mighty straight-eight

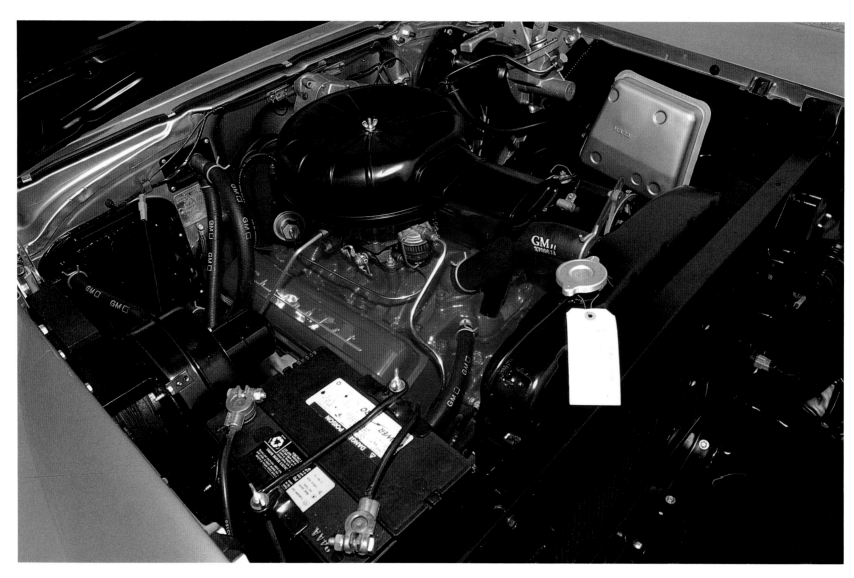

The 1957 Chevrolet received a completely new dashboard. The steering wheel was also redesigned. The interior fabric chosen was Silver vinyl and Black and Silver pattern cloth.

Duesenbergs, V-12 Packards, and V-16 Cadillacs were the epitome of style, elegance, and speed. Racing in the fifties provided a venue for this pent-up desire for performance. And enthusiasts did not need advertising to tell them what to do with their small-block-powered Chevys. The 283ci V-8 in the Chevrolet was a relative bargain, and the engine itself was supremely easy to hot rod.

What kind of performance could the top carbureted 283ci V-8 offer? *Motor Trend* tested a 1957 Bel Air, weighing roughly 3,500lb. Its dual four-barrel carbureted 283 offered 270hp and propelled the car from 0 to 60mph in 9.9sec and through the quarter-mile in 17.5sec. Chevrolet's nearest rivals included the 245hp Ford Fairlane (0–60mph in 11.1sec, 18.2sec quarter mile) and the 290hp Plymouth Belvedere convertible (0–60mph in 11.5sec, 17.7sec quarter mile).

The small-block V-8 Chevrolet did surprisingly well against its more-expensive and larger-displacement competition, staying within a few tenths of a second in both 0–60mph times and the quarter-mile. For example, the 325hp 392ci Hemi-equipped Chrysler New Yorker reached 60mph in 9.8sec and covered the quarter-mile in 17.3sec. In terms of performance for your money, the Chevrolet was unbeatable.

For comparison, *Road & Track* managed to get its hands on a 283hp 1957 fuel-injected Corvette, weighing under 3,000lb and equipped with the long-awaited four-speed manual transmission and Positraction rear axle with 4.11:1 gearing. It was truly a high-performance setup. The car rocketed to 60mph in 5.7sec, covered the quarter-mile in 14.3sec, and achieved a top speed of 132mph. The magazine wrote glowingly of the fuel-injected engine's performance in the Corvette, and said one could expect the same performance in the Chevrolet coupes and sedans: "The fuel-injected engine is an absolute jewel, quiet and remarkably docile when driven gently around town, yet instantly transformable into a roaring brute when pushed hard. It idles at about 900rpm and pulls easily and smoothly from this speed even in high gear. Its best feature is its instantaneous throttle response, completely free of any stutter or stumble under any situation. The

Previous page
This view from the passenger side shows just how voluminous the 1957 Chevrolet interior was. Men wore hats during the fifties and the cars of the day had to accommodate them.

The Chevrolet received all-new sheet metal for the 1957 model year. The fin era was in full bloom, and the 1957 Chevrolet embraced the theme. Chevrolet continued to emphasize the output of its small-block V-8, now up to 283ci and one horsepower per cubic inch. For 1957, Chevrolet was "Sweet, Smooth, and Sassy . . ."

The Bel Air Sport Coupe shows the solid quality of its Body by Fisher.

It likes to flex those big new muscles!

The Sweet, Smooth and Sassy '57 Chevrolet . . .

New muscles under the hood — with a choice of *five* precision-balanced new powerplants — to move you along in eager and effortless smoothness. New muscles to grip the road even more tightly and lay into the curves even more solidly. New muscles to give you a new lift in driving!

Here's a car designed to put the sparkle back into driving, a car that gives you that glad-to-be-alive feeling the moment you nudge the throttle! Part of the pleasure is performance—and part is the

wonderful sense of security that comes from Chevy's superb road-holding and precision control. It's a honey to handle on city streets, super-highways and everything in between.

How do you like to drive? There's a Chevy combination to suit every motoring mood, from the thrifty Six to the terrific Corvette V8, from the sports car *close-ratio* stick shift to the free-flight feeling of either of Chevrolet's two automatic drives.* Whenever the miles seem dull and motoring a chore, remember this: There's a sure cure just as close as your nearest Chevrolet dealer! . . . Chevrolet Division of General Motors, Detroit 2, Michigan.

*Corvette engine, close-ratio transmission, Powerglide and Turboglide automatic transmissions—optional at extra cost.

141

The Bel Air convertible enjoyed its best sales year in 1957, with 47,562 sold. Equipped with the V-8 and having a list price of $2,611, it was a most affordable way to experience the wind in your hair.

throttle linkage has a certain amount of backlash and friction, but there are no flat spots such as we described in last year's test of two Corvettes with twin four-barrel carburetors."

Zora Arkus-Duntov relished the displacement increase from 265ci to 283ci and was constantly seeking ways of boosting the small-block's perfor-

mance. He would sometimes appear in the dyno labs late at night so as not to arouse undue attention. Much of this performance development was directed toward the Corvette, but the results of that performance development filtered down to the passenger cars as well.

"One night when I was on second shift,"

smooth as quicksilver....

and quick as they come!
The new 1957 Chevrolet!

Trips never seemed so short — or roads so smooth — as they do in this beautiful traveler. It's sweet, smooth and sassy — with new velvety V8 power and a sure-footed way of going that's Chevy's alone!

SWEET, SMOOTH AND SASSY! The Bel Air 4-Door Sedan with Body by Fisher—one of 20 new Chevies.

Cars with Chevy's born-to-the-road build are hard to come by these days. In addition to designing a car that's just plain good to look at, Chevrolet engineers didn't forget that a car's first obligation is to ride, and ride *well*.

The new Chevy's low and wide (as comfortable inside as you'd want) and built for the road, with broad-based outrigger rear springs and beautifully balanced weight. That's the reason for Chevrolet's nice solid feel on the highway; the reason it clings to the road with such easy grace on curves.

Cars with Chevy's jack-rabbity pep are hard to come by, too. The secret here is a V8 engine, with up to 245 h.p.,* that brings a smile to your face when you nudge it even a little bit. Now if you're looking for extra economy in your driving, your choice would be Chevy's famous "Blue-Flame" Six. But V8 or 6, you're sure of fun. Stop by your Chevrolet dealer's and try one soon! . . . Chevrolet Division of General Motors, Detroit 2, Michigan.

*270-h.p. high-performance V8 engine also available at extra cost.

CHEVROLET

USA
57 CHEVROLET

The four-door Bel Airs were good sellers for Chevrolet in 1957. The four-door Bel Air sedan sold more than 260,000 units, while the Bel Air four-door sport sedan was purchased by nearly 143,000 buyers.

Painted Matador Red, Kevin Mueller's Bel Air convertible is every bit the head-turner today as the day it was driven out of the dealer's lot.

Next page
With top up and Glastron boat in tow, this was the good life for thousands of Americans in the late fifties. Yes, even the boats of the period had fins!

Mueller's Bel Air convertible lacks nothing in the power department. The 283ci 2x4-bbl Turbo-Fire V-8 ensured few would successfully challenge it on the open road.

Sherman remembered, "I was doing fuel economy curves, which are very mild tests. Mr. Duntov walked in and inquired as to what I was doing. I told him, and he said, 'I have an engine out there for a Corvette that I want tested. Why don't you remove this engine and put that one on?' I told him I'd be very pleased to do that, provided it was OK with supervision. I could not act on my own to interrupt the test. He said 'We'll take care of that little problem. Come on with me.'

"I shut the engine down and walked into the dyno control office and he asked the supervisor to please take that engine off and put the Corvette engine on. This was a 283 he wanted put on. The supervisor told him, 'No, I can't do that. I have this test to run.' Duntov said, 'Just a minute,' picked up the telephone, dialed a phone number, and handed the supervisor the telephone. I heard the supervisor say, 'Yes, Mr. Cole . . . yes, Mr. Cole.' Needless to say, in the next forty minutes, the engine I was testing was removed from the stand and this Corvette engine put on. He sat in the dynamometer cell while I ran the high output test on it for maximum horsepower. I believe that night we pulled something like 289 horsepower at 6200rpm with a single four-barrel carburetor and his designed camshaft. He was very famous for his cam profiles.

Above
The Bel Air Sport Sedan was introduced in 1956. Chevrolet called it "A beautiful new honey of a hardtop—with open sides, steel-top protection, and four-door convenience." It was continued for the 1957 model year.

Right
In 1957, the Bel Air Sport Sedan was Chevrolet's second most popular Bel Air model, selling 109,261 units. It was surpassed only by the Bel Air Sedan four-door which sold more than 282,000 units.

Next page
There are not many survivors of the 1957 Bel Air Sport Sedan today. This beautiful example finished in Sierra Gold with Adobe Biege is owned by Jim and Carolyn McLellan of Lakeland, Florida.

Top left
What made *The Black Widow* such a formidable competitor in its day was the 283hp Ramjet Fuel Injected Turbo-Fire V-8. Option no. RPO 578 cost $550. Production was limited and the option was difficult to get, despite many eager, prospective buyers.

Left
The very special engine under the hood was identified with crossed flags and "Fuel Injection" script on both sides of the car.

Above
The headlights were removed and these covers were mounted in their place.

Previous page
One of the finest surviving One-Fifty series Chevrolets in America is this 1957 two-door sedan. At 3,211lb, only the two-door Utility Sedan was lighter (by roughly 50lb), so the two-door Sedan served as an ideal base for a race car. This particular car became known in racing circles during the late fifties as *The Black Widow*.

"At the time," Sherman continued, "we didn't have computers, so as we were doing the tests, he was compiling the horsepower and plotting it on a curve sheet. We had to go many rpms over the horsepower peak. I got up to 6400rpm and I asked him, 'What do we look like?' He said, 'You won't believe what I'm going to tell you. Shut her down.' So I throttled it back. He said, 'Look at that.' And I said, 'Are you sure you calculated it right?' He said, 'Trust me.' We logged over one horsepower per cubic inch. That engine was re-

The two-door Sedan's center pillar gave it an inherently stronger upper structure—handy in the event of a rollover. To give the car some contrast, it was ordered in India Ivory and Onyx Black. The Black Widow is owned by Floyd Garrett of Fernandina Beach, Florida.

moved from the stand, drained, and shipped off to a car race. I believe it went down to Daytona."

The AMA ban on factory-sponsored racing in mid-1957 had a very desirable effect on Chevrolet's racing fortunes. Ford's chief executive officer, Robert McNamara, enforced the ban by stopping high-performance development and parts produc-

tion as well as sponsorship. Chevrolet, however, continued to furtively support the drivers and their teams. With no factory support after the AMA ban, the dominance of Fords in auto racing started to diminish. During the 1957 NASCAR season prior to the ban, Fords had won fifteen Grand National races and Chevrolets only five. After the ban

went into affect, Chevrolets won fourteen of the remaining thirty-two races and Fords won twelve. Buck Baker won the Grand National title for the second year in a row behind the wheel of a Chevrolet. Much the same happened in the NASCAR Convertible Division. Even though Fords and Mercurys won the majority of races during the year, the total point standings of the cars finishing races gave victory to Chevrolet, with Bob Welborn as the Division champion. Completing the triple crown of NASCAR events was the Short Track Division; although Ford squeaked by Chevrolet with fourteen victories to Chevrolet's thirteen, again, total points gave ultimate victory to Chevrolet.

Chevrolet also fared well in the 1957 sales race. Once again, Chevrolet was the leader, but by only a hair. Chevrolet sold 1,522,536 cars—only 130 more than Ford! This prompted a raging debate, which continues even to this day. Depending on how the numbers were run, Ford could also lay claim to the 1957 sales crown.

Other car makers of the day could only look on in envy.

Plymouth sold 655,526 cars, Buick sold 407,271 cars, Oldsmobile sold 390,091 cars, Pontiac followed with 343,298 cars, and Dodge sold 292,386. Ford's Mercury Division sold 274,820 cars, Cadillac sold 153,236 cars, Chrysler sold 118,733 cars, and DeSoto sold 117,747. American Motors managed to sell 114,083 cars in 1957. The merger of Studebaker and Packard did not stem the declining fortunes of these once-proud makes; combined, they sold only 72,889 cars, roughly one-third the number of cars these two makes had sold separately two years earlier. The new, highly touted but doomed Edsel sold 54,607 cars, followed by Imperial with 37,946 and Lincoln-Continental with 37,870.

A New Direction, New Cars, and the Tri-Chevy Legacy

Even as buyers were looking for end-of-the-year bargains in the fall of 1957, the all-new 1958 models began appearing in the Chevrolet showrooms. The 1958 was as radically different from the 1957 Chevrolet as the 1955 model was from the 1954. Unfortunately, the country was entering

The Black Widow was raced, naturally, without the wheel covers that originally came with the car.

a recession, and sales of all makes suffered considerably in 1958. Despite this, Chevrolet was again the sales leader, with 1,255,935 cars sold to Ford's 1,038,560. In 1959, Chevrolet introduced perhaps its most controversial model, stylistically speaking, and sales suffered as a result. Despite the fact that Chevrolet sold 1,428,962 cars in 1959, Ford sold 1,528,592. Champaign corks flew at Ford's Dearborn headquarters when the sales figures came in!

By 1960, the country was on its way to a booming economic recovery. Confident buyers descended on showrooms in record numbers that year. Chevrolet sold a re-cord 1,873,598 cars; Ford trailed with 1,511,504.

The 1955, 1956, and 1957 Chevrolets were soon forgotten in the industry-wide push for longer, lower, and wider. The horsepower race was also on and engine displacements began to grow. The sixties became one of the most exciting and dynamic decades in this nation's history.

The car that revolutionized Chevrolet—the

The interior of *The Black Widow* is factory stock, with the addition of a three-point seatbelt and required rollbar.

Tri-Chevy Accessories

The long list of factory options and dealer-installed accessories served two purposes on the classic Tri-Chevys: It allowed Chevrolet to minimize the base price of all models, thus maintaining its position as "the leader in the low-priced field," and it allowed a buyer to tailor a car to his needs and desires.

The many practical factory options included power steering and power brakes. The most expensive option, at one-fourth the list price of the car, was air conditioning. The dash-mounted autotronic eye, which automatically dimmed your high beams, was a feature one expected to see only in Cadillacs. And the Continental tire kit not only freed up room in the trunk, it gave the Chevrolet a distinctive look reminiscent of the classic cars of the thirties.

There were power accessories, too. The power front seat and power windows were luxury options that proved popular with Chevrolet buyers. There were also whimsical accessories, like the dash-mounted compass for those who wished to monitor their bearings. Police departments could order either driver-or passenger-side searchlights in order to check out late-night prowlers—or young couples parked where they shouldn't be.

Finally, there were numerous dress-up accessories like door handle guards and front fender ornaments, all designed to give your Chevrolet just a bit more flash. With so many factory-installed options, dealer-installed accessories, and a rainbow of exterior and interior colors to choose from, the buyer of a 1955, 1956, or 1957 Chevrolet could truly create a one-of-a-kind automobile.

For the directionally impaired, a dash-mounted compass.

Left to right: prismatic stoplight viewer, autotronic eye high beam dimmer, and searchlight.

Power driver's seat. Right, In case you couldn't recall what type of brakes you had . . .

Power window controls.

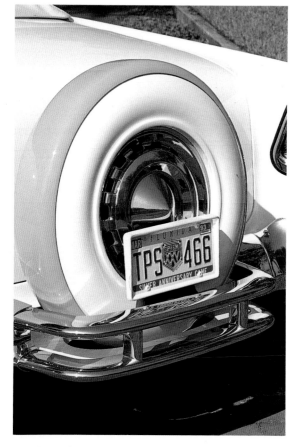

Continental kit lent class.

Even in the event of an untimely shower, the all-vinyl seats in Bel Air convertibles could be dried quickly. The Jacobsens' 1957 convertible is shown upholstered in Red Vinyl and Silver Pattern Vinyl.

Below and next page
Adding to the collectability of their Bel Air convertible, the Jacobsens' can point with pride to the Ramjet Fuel Injection Turbo-Fire V-8 under the hood.

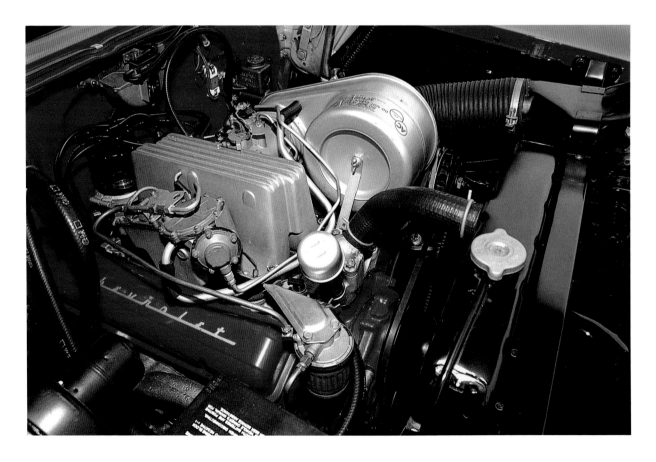

1955—became a cult favorite for hot rodders as its handsome lines refused to age. The 1956 and 1957 models were cast from the same die, so to speak, and the three model years became grouped together in the minds of Chevrolet enthusiasts. This following grew during the seventies and eighties and clubs were formed to embrace this enduring trio of cars. They became the object of countless restorations and customizing projects. It is safe to say millions of dollars have been expended on these three model years. These cars enjoy an undying devotion no other make can match. Can Ford or Chrylser enthusiasts boast their 1955, 1956, or 1957 models share the same status?

The classic Tri-Chevys have achieved their popularity as a result of wonderful styling, that superb small-block V-8 engine, and undeniable charisma.

These cars truly are greater than the sum of their parts. The fascination with the classic Tri-Chevys will not only endure, it will most assuredly grow in the decades to come.

With the 1958 model year came a completely new Chevrolet, and a completely new Bel Air. These 1958 and 1957 Bel Air convertibles are owned by Bill and Barbara Jacobsen.

No sooner had the clay bucks of the 1957 model been finalized then work was begun on what would be the 1958 Chevrolet—a model as completely re-engineered and redesigned as the 1955 model had been. As the showroom brochure stated: "No car in Chevrolet's field brings you more deep-down newness . . . with big, new changes from road to roof!"

To begin with, a new, larger displacement engine was planned for the car in order for it to compete with the larger and more powerful cars on the market and planned for the future. That engine was the 348ci Turbo-Thrust V-8, with its unique cylinder head and engine block design. It was designed to have more torque than the small-block V-8, though the trusty 283ci V-8 would be the midsize engine offering. The Blue-Flame 6 would continue as the economy engine.

Stylistically, the 1958 Chevrolet had little in common with the 1957 model. The front grille and bumper bore a family resemblance to the '57, but

there the similarities ended. Chevrolet moved from two to four headlights. At the other end of the car, the formerly vertical fins began the transition to horizontal ending up at a 45 degree angle.

Chevrolet was not going to be left behind the industry trend toward longer, lower, and wider cars. It abandoned the trim lines and compact dimen-sions of the 1955–1957 models. The car that emerged from the styling studio was laid over a 117.4in wheelbase—2.5in longer than before. Overall length grew from 200in to 209.1in. Width grew from 73.9in to 77.7in. Height for the hardtop models dropped from 58.5in to 57.2in.

Unadorned, it was amazing how bland the 1958 Chevrolet could be. The number desig-nations—One-Fifty and Two-Ten—were dropped in favor of Del Ray as the base trim level and Biscayne as the mid-level trim offering. However with the Bel Air, the chrome trim was tacked on in abundance, and the car trans-formed to one that looked like a luxury vehicle. Plus there was a new model within the Bel Air line—the Im-

pala—which offered even more glitz. With the Bel Air in general and the Impala in particular, wherever brightwork could be added it was. The Impala received, on top of the Bel Air trim, broad, ribbed lower sill panels; large, dummy four-vane vents in front of the rear wheels; and Impala script, animal logo, and crossed flags above the side vents and beneath the leading edge of the shallow fins of the rear fenders.

Some would argue that the 1958 Bel Air Impala had no more brightwork than the 1957 Bel Air, but the car's greater girth and the addition of the Impala trim level made the car more ostenta-tious in the eyes of others. It certainly was a radical departure from what Tri-Chevy enthusiasts later came to know and love.

Surprisingly, price was not up dramatically. The least-expensive model, the two-door sedan, was only $2,101. At the other end of the Chevrolet line was the Bel Air Impala convertible listing for $2,728. And what of the Nomad? Well, for 1958 it was made a model option and appeared as a four-door station wagon—a Nomad no more.

The automotive market in 1958 felt the new Chevrolet was the right car at the right time, and the car's look appealed to the car-buying public's aesthetic. The recession that began late in 1957 deepened by 1958, and total Chevrolet sales for 1958 were down by roughly 300,00 units. Had there not been a new model to draw prospective buyers to Chevrolet showrooms, sales undoubtedly would have declined even more dramatically.

The 1958 Chevrolet made it easier for the 1955, 1956, and 1957 models to be grouped together. The 1958 model year brought one design and engineering era to a close, and set another off in a new direction. Proof of Ed Cole's engineering wisdom was the demise of the 348 and later the demise of the 409; the small-block V-8, on the other hand, which began production in 1955, is still in production today powering a new generation of Chevrolets into the twenty-first century.

The 1958 Chevrolet marked the transition from vertical to horizontal fins, and a trend toward longer, lower, and wider.

This page and next page Hertz Rent a car featured Chevrolets in its ads in 1957. One ad conveys the fly-and-drive idea for the traveling businessman. The rate for a Chevrolet Bel Air back then was $7.55 a day and 8 cents a mile. The weekly rate was $38.50 plus 8 cents a mile. Of course, you could rent other General Motors cars from Hertz, too.

Today's luxurious planes provide unmatched speed and comfort.

Fly...get there faster! (He's got The Hertz Idea)

Pretty soft, eh? Flying rested and relaxed on a trip instead of putting up with tense hours driving on crowded highways. Well, why not! All you do is leave your car at home, take a fast plane, rent a Hertz car at your destination.

That's The Hertz Idea! You'll make more business calls and save time by always having a car to drive as your own. On your vacation, too, you'll enjoy more days of fun. A fast call to Hertz or any airline ticket office will reserve

a sparkling, new Powerglide Chevrolet Bel Air (or other fine car) in any of more than 700 cities. Your driver's license and proper identification give you the keys.

Drive wherever you like, as long as you like. The low national average rate is only $7.55 a day plus 8 cents per mile (lower by the week). That includes all gasoline, oil—even proper insurance. So next time—or *anytime*—you need a car, call Hertz. Look under "H" in

phone books everywhere (over 1,000 offices to serve you). Good idea! Hertz Rent A Car, 218 South Wabash Avenue, Chicago 4, Illinois.

More people by far...use
HERTZ
Rent a car

Hertz has new Powerglide Chevrolet Bel Airs or other fine cars at the Greater Fort Worth, Texas International Airport (Amon Carter Field).

CADILLAC

BUICK

OLDSMOBILE

CHEVROLET

Hertz rents the kind of cars you like to drive!

What's your pleasure? A Cadillac, maybe? Hertz rents Cadillacs. Hertz rents big Buicks and Oldsmobiles, too. Thousands of new Chevrolets and other fine cars. Station wagons, convertibles, sports cars. Take your pick at most Hertz offices.

They're all in A-1 condition, expertly maintained, more dependable, cleaner cars. More with power steering. That's The Hertz Idea. You'll get the kind of car you like to drive at

over 1,350 Hertz offices in more than 900 cities —world-wide. That's *more* offices by far where you can *rent* a car. *More* cities by far where you can *leave* a car. *More* locations where you can make a *reservation* for a car!

Just show your driver's license and proper identification. The national average rate for a new Powerglide Chevrolet Bel Air is only $38.50 a week plus 8 cents a mile. And that includes the cost of *all* the gasoline and oil you use en

route...and proper insurance. In addition to the Hertz charge card, we honor all air, rail, Diners' Club and hotel credit cards.

To be sure of a car at your destination—anywhere—use Hertz' more efficient reservation service. Call your local Hertz office for fast, courteous service. We're listed under "Hertz" in *alphabetical* phone books everywhere! Hertz Rent A Car, 218 South Wabash Avenue, Chicago 4, Illinois.

More people by far...use

HERTZ
Rent a car

"Rent it here...Leave it there" Now, nation-wide at no extra charge! (on rentals of $25.00 or more).

Index